# Technology Forecasting

# Contents

# Chapter 1

# Technology forecasting

**Technology forecasting** attempts to predict the future characteristics of useful technological machines, procedures or techniques.

## 1.1 Important aspects

Primarily, a technological forecast deals with the characteristics of technology, such as levels of technical performance, like speed of a military aircraft, the power in watts of a particular future engine, the accuracy or precision of a measuring instrument, the number of transistors in a chip in the year 2015, etc. The forecast does not have to state how these characteristics will be achieved.

Secondly, technological forecasting usually deals with only useful machines, procedures or techniques. This is to exclude from the domain of technological forecasting those commodities, services or techniques intended for luxury or amusement.

## 1.2 Rational and explicit methods

The whole purpose of the recitation of alternatives, is to show that there really is no alternative to forecasting. If a decisionmaker has several alternatives open to him, he will choose among them on the basis of which provides him with the most desirable outcome. Thus his decision is inevitably based on a forecast. His only choice is whether the forecast is obtained by rational and explicit methods, or by intuitive means.

The virtues of the use of rational methods are as follows:

1. They can be taught and learned,

2. They can be described and explained,

3. They provide a procedure followable by anyone who has absorbed the necessary training, and in some cases,

4. These methods are even guaranteed to produce the same forecast regardless of who uses them.

The virtue of the use of explicit methods is that they can be reviewed by others, and can be checked for consistency. Furthermore, the forecast can be reviewed at any subsequent time. Technology forecasting is not imagination.

## 1.3   Methods of technology forecasting

Commonly adopted methods of technology forecasting include the Delphi method, forecast by analogy, growth curves and extrapolation. Normative methods of technology forecasting — like the relevance trees, morphological models, and mission flow diagrams — are also commonly used.

THE DELPHI METHOD: The Delphi method is a structured communication technique, originally developed as a systematic, interactive forecasting method which relies on a panel of experts. In the standard version, the experts answer questionnaires in two or more rounds. After each round, a facilitator provides an anonymous summary of the experts' forecasts from the previous round as well as the reasons they provided for their judgments. Thus, experts are encouraged to revise their earlier answers in light of the replies of other members of their panel. It is believed that during this process the range of the answers will decrease and the group will converge towards the "correct" answer. Finally, the process is stopped after a pre-defined stop criterion (e.g. number of rounds, achievement of consensus, stability of results) and the mean or median scores of the final rounds determine the results.

## 1.4   Combining forecasts

Studies of past forecasts have shown that one of the most frequent reasons why a forecast goes wrong is that the forecaster ignores related fields.

A given technical approach may fail to achieve the level of capability forecast for it, because it is superseded by another technical approach which the forecaster ignored.

Another problem is that of inconsistency between forecasts. Because of these problems, it is often necessary to combine forecasts of different technologies. Therefore rather than to try to select the one method which is most appropriate, it may be better to try to combine the forecasts obtained by different methods.

If this is done, the strengths of one method may help compensate for the weaknesses of another.

### 1.4.1   Reasons for combining forecasts

The primary reason for combining forecasts of the same technology is to attempt to offset the weaknesses of one forecasting method with the strengths of another. In addition, the use of more than one forecasting method often gives the forecaster more insight into the processes at work which are responsible for the growth of the technology being forecast.

### 1.4.2   Trend curve and growth curves

A frequently used combination is that of growth curves and a trend curve for some technology. Here we see a succession of growth curves, each describing the level of functional capability achieved by a specific technical approach.

An overall trend curve is also shown, fitted to those items of historical data which represent the currently superior approach.

The use of growth curves and a trend curve in combination allows the forecaster to draw some conclusions about the future growth of a technology which might not be possible, were either method used alone.

With growth curves alone, the forecaster could not say anything about the time at which a given technical approach is likely to be supplanted by a successor approach.

With the trend curve alone, the forecaster could not say anything about the ability of a specific technical approach to meet the projected trend, or about the need to look for a successor approach. Thus the need for combining forecasts.

### 1.4.3   Identification of consistent deviations

Another frequently used combination of forecasts is that of the trend curve and one or more analogies.

We customarily consider the scatter of data points about a trend curve to be due to random influences which we can neither control nor even measure. However, consistent deviations may represent something other than just random influences.

Where such consistent deviations are identified, we may have an opportunity to apply an analogy. Typical events which bring about deviations from a trend are wars and depressions. Thus the purpose of combining analogies with a trend forecast is to predict deviations from the trend deviations which are associated with or caused by external events or influences.

As with other uses of analogy, it is important to determine the extent to which the analogy between the event used as the basis for the forecast, and the historical model event, satisfies the criteria for a valid analogy.

### 1.4.4 Forecasts of different technologies

Combining forecasts of different technologies may be even more important than combining the forecasts of the same technology.

One reason for this is the fact that technologies may interact or be interrelated in some fashion. Another reason for this is that of consistency in an overall picture or scenario. One of the simplest examples of interacting trends is the projection to absurdity, i.e. simply projecting the given data indefinitely without getting any specific result. For instance, if one simply projects recent rates of growth of world population, one arrives at some fantastic conclusions about the density of population in a particular place by various dates in the next millennium.

Some other trends which can confidently be expected to not continue indefinitely are:

1. Annual production of scientific papers.

2. Number of automobiles per capita.

3. Kilowatt hours of electricity generated annually.

Another instance of interacting trends was in the case of the number of scientists in the U.S. growing faster than the overall population. Since the 1940s through the 1960s, science as an activity in the United States grew exponentially. The number of dollars spent on R&D was growing faster than the GNP (in the 1960s).

If projected indefinitely, these two curves would give the result that eventually every person in the U.S. would be working as a scientist and the entire GNP would be devoted to R&D alone, which are however absurd conclusions. Thus it is clear that the scientific discipline of technology forecasting is not mere trend extrapolation but also involves combining forecasts.

## 1.5 Uses in manufacturing

Almost all modern manufacturing firms utilize the services of a technological forecaster. Nevertheless, there are a number of alternatives to the rational and explicit forecasting of technology, such as 'no forecast', 'anything can happen' (i.e. relying on pure chance), 'window-blind forecasting', 'genius forecasting' and boasting of a 'glorious past' (i.e. adopting the same old techniques).

Thus technological forecasting is not mere astrology or palmistry, but a scientific and well defined procedure adopted by a technological forecaster or a consultancy for the forecasting of a particular technology. Even though technological forecasting is a scientific discipline, some experts are of the view that "the only certainty of a particular forecast is that it is wrong to some degree."

## 1.6 Forecasting institutes

- TechCast Project

- Singularity Institute for Artificial Intelligence

- Future of Humanity Institute

- The Millennium Project

- Institute for the Future

## 1.7   See also

- Accelerating change

- Delphi method

- Forecasting

- Futurology

- List of emerging technologies

- Optimism bias

- Reference class forecasting

- Strategic Foresight

- Technology Scouting

- Technology roadmap

## 1.8   References

- Klopfenstein, Bruce K. "Forecasting consumer adoption of information technology and services - Lessons from home video forecasting". *Journal of the American Society for Information Science* 1989 Jan;40(1):17-26.

- Martino, Joseph (January 1983). *Technological Forecasting for Decision Making* (2nd ed.). North-Holland. ISBN 0-444-00722-9.

- Makridakis, Spyros; Steven C. Wheelwright; Rob J. Hyndman (December 1998). *Forecasting: Methods and Applications* (3rd ed.). John Wiley. ISBN 0-471-53233-9.

- Twiss, Brian C. (July 1, 1992). *Forecasting for Technologists and Engineers: A Practical Guide for Better Decisions*. Institution of Electrical Engineers. ISBN 0-86341-265-3.

## 1.9   External links

- TechCast Article Series, William Halal Next Next Things

- TSTC Forecasting The emerging technology & forecasting office at Texas State Technical College

- UNIDO TECHNOLOGY FORESIGHT MANUAL

# Chapter 2

# 2081: A Hopeful View of the Human Future

Princeton physicist Gerard K. O'Neill's 1981 book, *2081: A Hopeful View of the Human Future* was an attempt to predict the technological and social state of humanity 100 years in the future. O'Neill's positive attitude towards both technology and human potential distinguished this book from gloomy predictions of a Malthusian catastrophe by contemporary scientists. Paul R. Ehrlich wrote in 1968 in *The Population Bomb*, "in the 1970s and 1980s hundreds of millions of people will starve to death". The Club of Rome's 1972 *Limits to Growth* predicted a catastrophic end to the Industrial Revolution within 100 years from resource exhaustion and pollution.

O'Neill's contrary view had two main components. First, he analyzed the previous attempts to predict the future of society—including many catastrophes that had not materialized. Second, he extrapolated historical trends under the assumption that the obstacles identified by other authors would be overcome by five technological "Drivers of Change". He extrapolated an average American family income in 2081 of $1 million/year. Two developments based on his own research were responsible for much of his optimism. In *The High Frontier: Human Colonies in Space* O'Neill described solar power satellites that provide unlimited clean energy, making it far easier for humanity to reach and exceed present developed-world living standards. Overpopulation pressures would be relieved as billions of people eventually emigrate to colonies in free space. These colonies would offer an Earth-like environment but with vastly higher productivity for industry and agriculture. These colonies and satellites would be constructed from asteroid or lunar materials launched into the desired orbits cheaply by the mass drivers O'Neill's group developed.

## 2.1 Part I: The Art of Prophecy

Previous futurist authors he cites:

- Edward Bellamy

- J.D. Bernal

- McGeorge Bundy

- Arthur C. Clarke

- George Darwin

- J. B. S. Haldane

- Robert Heilbroner

- Aldous Huxley

- Rudyard Kipling

- Thomas More

- George Orwell

- George Thompson

- Konstantin Tsiolkovski

- Jules Verne

- H.G. Wells

- Eugene Zamiatin

Arthur C. Clarke's *Profiles of the Future* included a long list of predictions, many of which O'Neill endorsed. Two of his maxims that O'Neill quotes [1] seem to sum up O'Neill's attitude, as well:

- "anything that is theoretically possible will be achieved in practice, no matter what the technical difficulties, if it is desired greatly enough"

- "We can never run out of energy or matter, but we can all too easily run out of brains."

## 2.2   Part II: The Drivers of Change

Sections are included on the five key "Drivers of Change" believed by O'Neill to be the focus of future development:

- Automation

- Space Colonies

- Communications

- Computers

- Energy

O'Neill applied basic physics to understand the limits of possible change, using the history of the technology to extrapolate likely progress. He applied the history of computing to reason about how people and institutions will shape and be shaped by the likely changes. He predicted that future computers must run at a very low-voltage because of heat. The main basis of his technology extrapolation for computers is Moore's Law, one of the greatest successes of Trend estimation in predicting human progress.

His predicted the social aspects of the future of computers. He identified computers as the most certain of his five "drivers of change", because their adoption could be driven primarily by individual or local decisions, while the other four such as space colonies depended on large-scale decision-making. He observed the success of minicomputers, calculators, and the first home computers, and predicted that every home would have a computer in a hundred years. With the aid of speculations by computer pioneers such as John von Neumann and the writers of "tracts" such as Zamyatin's *We*, O'Neill also predicted that privacy would be under siege from computers in 2081.

O'Neill predicted that software engineering issues and the intractability of artificial intelligence problems would require massive programming efforts and very powerful processors to achieve truly usable computers. His prediction was based on the difficulties and failures of computer use he had observed in 1981, including a candid horror story of his own Princeton University library's attempt to computerize its operations. His computers of the future, represented by the robot butler his visitor to Earth encounters in 2081, included speaker-independent speech recognition and natural language processing. O'Neill correctly pointed out the huge difference between computers and human brains, and stated that, while a more human-like artificial brain is a worthy goal, computers will be vastly improved descendants of today's rather than truly intelligent and creative artificial brains.

## 2.3 Part III: The World in 2081

This section was written as a series of dispatches home from "Eric C. Rawson", a native of a distant space colony called "Fox Cluster". By analogy with American religious colonists such as the Puritans and Mormons, O'Neill suggests that such a colony might have been founded by a group of pacifists who chose to live about twice as far from the Sun as Pluto in order to avoid involvement in Earth's wars. His calculations indicate that colonies at this distance could have Earth-level sunlight using a mirror the same weight as the colony itself. Eric pays a visit to the Earth of 2081 to take care of family business and explore a world that is nearly as foreign to him as it is to us.

After each dispatch, O'Neill added a section that described his reasoning for each situation the visitor described, such as riding a "floater" train going thousands of miles per hour in vacuum, interacting with a household robot or visiting a fully enclosed Pennsylvania city with a tropical climate in midwinter. Each section was written from his perspective as a physicist. For example, his description of "Honolulu, Pennsylvania" included multiple roof layers that could be retracted in good weather. The city enjoyed an artificial tropical climate all year because of internal climate controls and advanced insulation. He also proposed magnetically levitated "floater" trains moving in very-low-pressure tunnels that would replace airplanes on heavily traveled routes.

## 2.4 Part IV: Wild Cards

This section explores, not the most probable outcomes, but "the limits of the possible": how likely some scenarios O'Neill considered less probable are, and what they might mean. These included nuclear annihilation, attaining immortality, and contact with extraterrestrial civilizations. For this last case, he presents a thought experiment about how a hypothetical alien civilization, the "Primans", could explore the galaxy with self-replicating robots, monitoring every planetary system in the Galaxy without betraying their own position, and destroying intelligent life (by building giant mirrors to incinerate the planet) if they felt threatened. This experiment seems to prove that conflict or even surprise contact with an intelligent alien life form—that staple of science fiction—is highly unlikely.

## 2.5 See also

- Orbiting skyhooks

### 2.5.1 Prediction

- Futures studies

- 2000s in science and technology

### 2.5.2 Technologies discussed

- Space advocacy

- Space technology

- Space colonization

- Solar power satellite

- Asteroid mining

- Space elevator

- Space manufacturing

- Space mining

- Space-based industry

- Domed city

## 2.6  References

[1]  O'Neill 1981: 27

- O'Neill, Gerard K. (1977). *The High Frontier: Human Colonies in Space*. William Morrow & Company. ISBN 978-0-9622379-0-4.

- O'Neill, Gerard K. (1977). *Space-Based Manufacturing from Nonterrestrial Materials*. Amer Inst of Aeronautics. ISBN 978-0-915928-21-7.

- O'Neill, Gerard K. (1981). *2081: A Hopeful View of the Human Future*. Simon and Schuster. ISBN 978-0-671-44751-9.

- NSS review of *2081*

# Chapter 3

# 5G

For other uses, see 5G (disambiguation).

**5G** (**5th generation mobile networks** or **5th generation wireless systems**) denotes the next major phase of mobile telecommunications standards beyond the current 4G/IMT-Advanced standards. 5G has speeds beyond what the current 4G can offer.

The Next Generation Mobile Networks Alliance defines the following requirements for 5G networks:

- Data rates of several tens of megabits per second should be supported for tens of thousands of users

- 1 gigabit per second to be offered simultaneously to tens of workers on the same office floor

- Several hundreds of thousands of simultaneous connections to be supported for massive sensor deployments

- Spectral efficiency should be significantly enhanced compared to 4G

- Coverage should be improved

- Signalling efficiency should be enhanced

- Latency should be reduced significantly compared to LTE[1]

The Next Generation Mobile Networks Alliance feels that 5G should be rolled out by 2020 to meet business and consumer demands.[2] In addition to providing simply faster speeds, they predict that 5G networks also will need to meet the needs of new use cases, such as the Internet of Things as well as broadcast-like services and lifeline communication in times of natural disaster.

Although updated standards that define capabilities beyond those defined in the current 4G standards are under consideration, those new capabilities are still being grouped under the current ITU-T 4G standards.

## 3.1   Background

A new mobile generation has appeared approximately every 10 years since the first 1G system, Nordic Mobile Telephone, was introduced in 1982. The first 2G system was commercially deployed in 1992, and the first 3G system appeared in 2001. 4G systems fully compliant with IMT Advanced were first standardized in 2012. The development of the 2G (GSM) and 3G (IMT-2000 and UMTS) standards took about 10 years from the official start of the R&D projects, and development of 4G systems began in 2001 or 2002.[3][4] Predecessor technologies have been present on the market a few years before the new mobile generation, for example the pre-3G system CdmaOne/IS95 in the US in 1995, and the pre-4G systems Mobile WiMAX in South-Korea 2006, and first release-LTE in Scandinavia 2009. In April 2008, NASA partnered with Machine-to-Machine Intelligence (M2Mi) Corp to develop 5G communication technology[5]

Mobile generations typically refer to non–backward-compatible cellular standards following requirements stated by ITU-R, such as IMT-2000 for 3G and IMT-Advanced for 4G. In parallel with the development of the ITU-R mobile generations, IEEE and other standardization bodies also develop wireless communication technologies, often for higher data rates and higher frequencies but shorter transmission ranges. The first gigabit IEEE standard was IEEE 802.11ac, commercially available since 2013, soon to be followed by the multigigabit standard WiGig or IEEE 802.11ad.

## 3.2   Debate

Based on the above observations, some sources suggest that a new generation of 5G standards may be introduced approximately in the early 2020s.[6][7] However, significant debate continued, on what 5G is about exactly. Prior to 2012, some industry representatives expressed skepticism toward 5G.[8] 3GPP held a conference in September 2015 to plan development of the new standard.[9]

New mobile generations are typically assigned new frequency bands and wider spectral bandwidth per frequency channel (1G up to 30 kHz, 2G up to 200 kHz, 3G up to 5 MHz, and 4G up to 20 MHz), but skeptics argue that there is little room for larger channel bandwidths and new frequency bands suitable for land-mobile radio.[8] The higher frequencies would overlap with aka band transmissions of communication satellites.[10] From users' point of view, previous mobile generations have implied substantial increase in peak bitrate (i.e. physical layer net bitrates for short-distance communication), up to 1 gigabit per second to be offered by 4G.

If 5G appears and reflects these prognoses, the major difference from a user point of view between 4G and 5G techniques must be something other than increased peak bit rate. For example, higher number of simultaneously connected devices, higher system spectral efficiency (data volume per area unit), lower battery consumption, lower outage probability (better coverage), high bit rates in larger portions of the coverage area, lower latencies, higher number of supported devices, lower infrastructure deployment costs, higher versatility and scalability, or higher reliability of communication. Those are the objectives in several of the research papers and projects below.

GSMHistory.com[11] has recorded three very distinct 5G network visions that had emerged by 2014:

**A super-efficient mobile network** that delivers a better performing network for lower investment cost. It addresses the mobile network operators' pressing need to see the unit cost of data transport falling at roughly the same rate as the volume of data demand is rising. It would be a leap forward in efficiency based on the IET Demand Attentive Network (DAN) philosophy.[12]

**A super-fast mobile network** comprising the next generation of small cells densely clustered to give a contiguous coverage over at least urban areas and getting the world to the final frontier of true "wide-area mobility." It would require access to spectrum under 4 GHz perhaps via the world's first global implementation of Dynamic Spectrum Access.

**A converged fiber-wireless network** that uses, for the first time for wireless Internet access, the millimeter wave bands (20 – 60 GHz) so as to allow very-wide-bandwidth radio channels able to support data-access speeds of up to 10 Gbit/s. The connection essentially comprises "short" wireless links on the end of local fiber optic cable. It would be more a "nomadic" service (like Wi-Fi) rather than a wide-area "mobile" service.

## 3.3   Research & Development projects

In 2008, the South Korean IT R&D program of "5G mobile communication systems based on beam-division multiple access and relays with group cooperation" was formed.[13]

In 2012, the UK Government announced the establishment of a 5G Innovation Centre at the University of Surrey – the world's first research center set up specifically for 5G mobile research.[14]

In 2012, NYU WIRELESS was established as a multidisciplinary research center, with a focus on 5G wireless research, as well as its use in the medical and computer-science fields. The center is funded by the National Science Foundation and a board of 10 major wireless companies (as of July 2014) that serve on the Industrial Affiliates board of the center. NYU WIRELESS has conducted and published channel measurements that show that millimeter wave frequencies will be viable for multigigabit-per-second data rates for future 5G networks.

In 2012, the European Commission, under the lead of Neelie Kroes, committed 50 million euros for research to deliver 5G mobile technology by 2020.[15] In particular, The METIS 2020 Project is driven by several telecommunication companies, and aims at reaching world-wide consensus on the future global mobile and wireless communication system. The METIS overall technical goal is to provide a system concept that supports 1000 times higher mobile system spectral efficiency, compared to current LTE deployments.[7] In addition, in 2013, another project has started, called 5GrEEn,[16] linked to project METIS and focusing on the design of green 5G mobile networks. Here the goal is to develop guidelines for the definition of a new-generation network with particular emphasis on energy efficiency, sustainability and affordability.

In November 2012, a research project funded by the European Union under the ICT Programme FP7 was launched under the coordination of IMDEA Networks Institute (Madrid, Spain): i-JOIN (Interworking and JOINt Design of an Open Access and Backhaul Network Architecture for Small Cells based on Cloud Networks). iJOIN introduces the novel concept of the radio access network (RAN) as a service (RANaaS), where RAN functionality is flexibly centralized through an open IT platform based on a cloud infrastructure. iJOIN aims for a joint design and optimization of access and backhaul, operation and management algorithms, and architectural elements, integrating small cells, heterogeneous backhaul and centralized processing. Additionally to the development of technology candidates across PHY, MAC, and the network layer, iJOIN will study the requirements, constraints and implications for existing mobile networks, specifically 3GPP LTE-A.

In January 2013, a new EU project named CROWD (Connectivity management for eneRgy Optimised Wireless Dense networks) was launched under the technical supervision of IMDEA Networks Institute, to design sustainable networking and software solutions for the deployment of very dense, heterogeneous wireless networks. The project targets sustainability targeted in terms of cost effectiveness and energy efficiency. Very high density means 1000x higher than current density (users per square meter). Heterogeneity involves multiple dimensions, from coverage radius to technologies (4G/LTE vs. Wi-Fi), to deployments (planned vs. unplanned distribution of radio base stations and hot spots).

In September 2013, the Cyber-Physical System (CPS) Lab at Rutgers University, NJ, started to work on dynamic provisioning and allocation under the emerging cloud radio-access network (C-RAN). They have shown that the dynamic demand-aware provisioning in the cloud will decrease the energy consumption while increasing the resource utilization.[17] They also have implemented a test bed for feasibility of C-RAN and developed new cloud-based techniques for interference cancellation. Their project is funded by the National Science Foundation.

In November 2013, Chinese telecom equipment vendor Huawei said it will invest $600 million in research for 5G technologies in the next five years.[18] The company's 5G research initiative does not include investment to productize 5G technologies for global telecom operators. Huawei will be testing 5G technology in Malta.[19][20]

In 2015 Huawei and Ericsson are testing 5G-related technologies in rural areas in northern Netherlands.[21]

In July 2015 the 5GNORMA project was launched. The key objective of 5G NORMA is to develop a conceptually novel, adaptive and future-proof 5G mobile network architecture. The architecture is enabling unprecedented levels of network customisability, ensuring stringent performance, security, cost and energy requirements to be met; as well as providing an API-driven architectural openness, fuelling economic growth through over-the-top innovation. With 5G NORMA, leading players in the mobile ecosystem aim to underpin Europe's leadership position in 5G.[22]

In July 2015 the European research project mmMAGIC was launched. The mmMAGIC project will develop new concepts for mobile radio access technology (RAT) for mmwave band deployment. This is a key component in the 5G multi-RAT ecosystem and will be used as a foundation for global standardization. The project will enable ultrafast mobile broadband services for mobile users, supporting UHD/3D streaming, immersive applications and ultra-responsive cloud services. A new radio interface, including novel network management functions and architecture components will be designed taking as guidance 5G PPP's KPI and exploiting the use of novel adaptive and cooperative beam-forming and tracking techniques to address the specific challenges of mm-wave mobile propagation. The ambition of the project is to pave the way for a European head start in 5G standards and to strengthen European competitiveness. The consortium brings together major infrastructure vendors, major European operators, leading research institutes and universities, measurement equipment vendors and one SME. [23]

In July 2015 IMDEA Networks launched the Xhaul project, as part of the European H2020 5G Public-Private Partnership

(5G PPP). Xhaul will develop an adaptive, sharable, cost-efficient 5G transport network solution integrating the fronthaul and backhaul segments of the network. This transport network will flexibly interconnect distributed 5G radio access and core network functions, hosted on in-network cloud nodes. Xhaul will greatly simplify network operations despite growing technological diversity. It will hence enable system-wide optimisation of Quality of Service (QoS) and energy usage as well as network-aware application development. The Xhaul consortium comprises 21 partners including leading telecom industry vendors, operators, IT companies, small and medium-sized enterprises and academic institutions. [24]

In July 2015 the European 5G research project Flex5Gware was launched. The objective of Flex5Gware is to deliver highly reconfigurable hardware (HW) platforms together with HW-agnostic software (SW) platforms targeting both network elements and devices and taking into account increased capacity, reduced energy footprint, as well as scalability and modularity, to enable a smooth transition from 4G mobile wireless systems to 5G. This will enable that 5G HW/SW platforms can meet the requirements imposed by the anticipated exponential growth in mobile data traffic (1000 fold increase) together with the large diversity of applications (from low bit-rate/power power for M2M to interactive and high resolution applications).[25]

## 3.4   Research

Key concepts suggested in scientific papers discussing 5G and beyond 4G wireless communications are:

The IEEE Journal on Selected Areas in Communications published a special issue on 5G. See the issue for June 2014, containing, among other papers, a comprehensive survey of 5G enabling technologies and solutions.[26] IEEE Spectrum has a story about millimeter-wave wireless communications as a viable means to support 5G in its September 2014 issue.

- Radio propagation and channel models for millimeter-wave wireless communication may be found in IEEE papers: Millimeter Wave Mobile Communications for 5G Cellular: It Will Work!" in IEEE Access, Vol. 1, May 2013; "Broadband Millimeter-Wave Propagation Measurements and Models Using Adaptive-Beam Antennas for Outdoor Urban Cellular Communications, in IEEE Trans. Antennas and Propagation, April 2013, and many other peer-reviewed conference and journal papers. Pearson/Prentice Hall has released a comprehensive text on "Millimeter Wave Wireless Communications," authored by Ted Rappaport, R. W Heath, Jr., Robert Daniels, and James Murdock. This text, over 700 pages in length, covers technical areas regarding potential 5G technologies, including standards for major global 60 GHz wireless local-area networks (WLAN) and personal local-area networks (WPAN).

- Massive Dense Networks also known as Massive Distributed MIMO providing green flexible small cells 5G Green Dense Small Cells. This is a transmission point equipped with a very large number of antennas that simultaneously serve multiple users. With massive MIMO multiple messages for several terminals can be transmitted on the same time-frequency resource, maximizing beamforming gain while minimizing interference.[27][28][29][30]

- Advanced interference and mobility management, achieved with the cooperation of different transmission points with overlapped coverage, and encompassing the option of a flexible use of resources for uplink and downlink transmission in each cell, the option of direct device-to-device transmission and advanced interference cancellation techniques.[31][32][33]

- Efficient support of machine-type devices to enable the Internet of Things with potentially higher numbers of connected devices, as well as novel applications, such as mission-critical control or traffic safety, requiring reduced latency and enhanced reliability.

- Use of millimeter-wave frequencies (e.g. up to 90 GHz) for wireless backhaul and/or access (IEEE rather than ITU generations)

- Pervasive networks providing Internet of things, wireless sensor networks and *ubiquitous computing*: The user can be connected simultaneously to several wireless access technologies and can move seamlessly between them (See Media independent handover or vertical handover, IEEE 802.21, also expected to be provided by future 4G releases. See also multihoming.). These access technologies can be 2.5G, 3G, 4G, or 5G mobile networks, Wi-Fi, WPAN, or any other future access technology. In 5G, the concept may be further developed into multiple concurrent data-transfer paths.[34]

- Multiple-hop networks: A major issue in systems beyond 4G is to make the high bit rates available in a larger portion of the cell, especially to users in an exposed position in between several base stations. In current research, this issue is addressed by cellular repeaters and macro-diversity techniques, also known as group cooperative relay, where users also could be potential cooperative nodes, thanks to the use of direct device-to-device (D2D) communication.[13]

- Wireless network virtualization: Virtualization will be extended to 5G mobile wireless networks. With wireless network virtualization, network infrastructure can be decoupled from the services that it provides, where differentiated services can coexist on the same infrastructure, maximizing its utilization. Consequently, multiple wireless virtual networks operated by different service providers (SPs) can dynamically share the physical substrate wireless networks operated by mobile network operators (MNOs). Since wireless network virtualization enables the sharing of infrastructure and radio spectrum resources, the capital expenses (CapEx) and operation expenses (OpEx) of wireless (radio) access networks (RANs), as well as core networks (CNs), can be reduced significantly. Moreover, mobile virtual network operators (MVNOs) who may provide some specific telecom services (e.g., VoIP, video call, over-the-top services) can help MNOs attract more users, while MNOs can produce more revenue by leasing the isolated virtualized networks to them and evaluating some new services.[35]

- Cognitive radio technology, also known as smart radio. This allows different radio technologies to share the same spectrum efficiently by adaptively finding unused spectrum and adapting the transmission scheme to the requirements of the technologies currently sharing the spectrum. This dynamic radio resource management is achieved in a distributed fashion and relies on software-defined radio.[36][37] See also the IEEE 802.22 standard for Wireless Regional Area Networks.

- Dynamic Adhoc Wireless Networks (DAWN),[3] essentially identical to Mobile ad hoc network (MANET), Wireless mesh network (WMN) or wireless grids, combined with smart antennas, cooperative diversity and flexible modulation.

- Vandermonde-subspace frequency division multiplexing (VFDM): a modulation scheme to allow the co-existence of macro cells and cognitive radio small cells in a two-tiered LTE/4G network.[38]

- IPv6, where a visiting care-of mobile IP address is assigned according to location and connected network.[34]

- Wearable devices with AI capabilities.[3] such as smartwatches and optical head-mounted displays for augmented reality

- One unified global standard.[3]

- *Real wireless world* with no more limitation with access and zone issues.[34]

- *User centric* (or *cell phone developer initiated*) network concept instead of operator-initiated (as in 1G) or system developer initiated (as in 2G, 3G and 4G) standards[39]

- Li-Fi (a portmanteau of *light* and *Wi-Fi*) is a massive MIMO visible light communication network to advance 5G. Li-Fi uses light-emitting diodes to transmit data, rather than radio waves like Wi-Fi.[40]

- *Worldwide wireless web* (WWWW), i.e. comprehensive wireless-based web applications that include full multimedia capability beyond 4G speeds.[3]

## 3.5 History

- In April 2008, NASA partnered with Geoff Brown and Machine-to-Machine Intelligence (M2Mi) Corp to develop 5G communication technology[5]

- In 2008, the South Korean IT R&D program of "5G mobile communication systems based on beam-division multiple access and relays with group cooperation" was formed.[13]

- On 8 October 2012, the UK's University of Surrey secured £35M for a new 5G research center, jointly funded by the British government's UK Research Partnership Investment Fund (UKRPIF) and a consortium of key international mobile operators and infrastructure providers, including Huawei, Samsung, Telefonica Europe, Fujitsu Laboratories Europe, Rohde & Schwarz, and Aircom International. It will offer testing facilities to mobile operators keen to develop a mobile standard that uses less energy and less radio spectrum while delivering speeds faster than current 4G with aspirations for the new technology to be ready within a decade.[41][42][43][44]

- On 1 November 2012, the EU project "Mobile and wireless communications Enablers for the Twenty-twenty Information Society" (METIS) starts its activity towards the definition of 5G. METIS intends to ensure an early global consensus on these systems. In this sense, METIS will play an important role of building consensus among other external major stakeholders prior to global standardization activities. This will be done by initiating and addressing work in relevant global fora (e.g. ITU-R), as well as in national and regional regulatory bodies.[45]

- Also on November 2012, the iJOIN EU project was launched, focusing on "small cell" technology, which is of key importance for taking advantage of limited and strategic resources, such as the radio wave spectrum. According to Günther Oettinger, the European Commissioner for Digital Economy and Society (2014–19), "an innovative utilization of spectrum" is one of the key factors at the heart of 5G success. Oettinger further described it as "the essential resource for the wireless connectivity of which 5G will be the main driver".[46] iJOIN was selected by the European Commission as one of the pioneering 5G research projects to showcase early results on this technology at the Mobile World Congress 2015 (Barcelona, Spain).

- In February 2013, ITU-R Working Party 5D (WP 5D) started two study items: (1) Study on IMT Vision for 2020 and beyond, and; (2) Study on future technology trends for terrestrial IMT systems. Both aiming at having a better understanding of future technical aspects of mobile communications towards the definition of the next generation mobile.

- On 12 May 2013, Samsung Electronics stated that they have developed the world's first "5G" system. The core technology has a maximum speed of tens of Gbit/s (gigabits per second). In testing, the transfer speeds for the "5G" network sent data at 1.056 Gbit/s to a distance of up to 2 kilometres.with the use of an 8*8 MIMO.[47][48]

- In July 2013, India and Israel have agreed to work jointly on development of fifth generation (5G) telecom technologies.[49]

- On 1 October 2013, NTT (Nippon Telegraph and Telephone), the same company to launch world first 5G network in Japan, wins Minister of Internal Affairs and Communications Award at CEATEC for 5G R&D efforts[50]

- On 6 November 2013, Huawei announced plans to invest a minimum of $600 million into R&D for next generation 5G networks capable of speeds 100 times faster than modern LTE networks.[51]

- On 8 May 2014, NTT DoCoMo start testing 5G mobile networks with Alcatel Lucent, Ericsson, Fujitsu, NEC, Nokia and Samsung.[52]

- In June 2014, the EU research project CROWD was selected by the European Commission to join the group of "early 5G precursor projects". These projects contribute to the early showcasing of potential technologies for the future ubiquitous, ultra-high bandwidth "5G" infrastructure. CROWD was included in the list of demonstrations at the European Conference on Networks and Communications (EuCNC) organized by the EC in June 2014 (Italy).

- At the end of September 2014, Dresden university inaugurates a 5G laboratory in partnership with Vodafone.[53]

- On October 2014, the research project TIGRE5-CM (Integrated technologies for management and operation of 5G networks) is launched with the aim to design an architecture for future generation mobile networks, based on the SDN (Software Defined Networking) paradigm. IMDEA Networks Institute is the project coordinator.

- In November 2014, it was announced that Megafon and Huawei will be developing a 5G network in Russia. A pilot network will be available by the end of 2017, just in time for the 2018 World Cup.[54][55]

- On 19 November 2014, Huawei and SingTel announced the signing of a MoU to launch a joint 5G innovation program.[56]

- On 28 April 2015, President Recep Tayyip Erdoğan announced Turkey might cancel 4G tender and move straight to 5G from 3G directly in two years.[57]

- On 22 June 2015, Greek government announced to Euro-group council talks that potential licensing 5G and 4G technology would offer 350 million euros earnings, as a result they were criticized for misleading European leaders in producing potential earnings from a technology that is supposed to roll-out after 2020.[58]

- On 8 September 2015, Verizon announced a roadmap to begin testing 5G in field trials in the United States in 2016.[59]

- On the 1st of October 2015, the French Operator Orange announced to be about to deploy 5G technologies to begin the first trial on January 2016 in Belfort, a City of Eastern France.[60]

## 3.6 See also

- List of mobile phone generations

- Femtocell

- IEEE 802.11u authentication

- IEEE P1905 hybrid networking

- Ka band

- OpenFlow/OpenRadio for sharing backhaul.

- Picocell

- Ultra-wideband (UWB)

## 3.7 References

[1] http://www.techrepublic.com/article/does-the-world-really-need-5g

[2] https://www.ngmn.org/uploads/media/NGMN_5G_White_Paper_V1_0.pdf

[3] Akhtar, Shakil (August 2008) [2005]. Pagani, Margherita, ed. *2G-5G Networks: Evolution of Technologies, Standards, and Deployment* (Second ed.). Hershey, Pennsylvania, United States: IGI Global. pp. 522–532. doi:10.4018/978-1-60566-014-1.ch070. ISBN 978-1-60566-014-1. Archived from the original (pdf) on 2 June 2011. Retrieved 2 June 2011.

[4] *Emerging Wireless Technologies; A look into the future of wireless communication – beyond 3G* (PDF). SafeCom (a US Department of Homeland Security program). Retrieved 27 September 2013. Since the general model of 10 years to develop a new mobile system is being followed, that time line would suggest 4G should be operational some time around 2011.

[5] "NASA Ames Partners With M2MI For Small Satellite Development".

[6] Xichun Li; Abudulla Gani; Rosli Salleh; Omar Zakaria (February 2009). "The Future of Mobile Wireless Communication Networks" (pdf). International Conference on Communication Software and Networks. ISBN 978-0-7695-3522-7. Retrieved 27 September 2013.

[7] "The METIS 2020 Project – Mobile and Wireless Communication Enablers for the 2020 Information Society" (pdf). METIS. 6 July 2013. Retrieved 27 September 2013.

[8] "Interview with Ericsson CTO: There will be no 5G - we have reached the channel limits". DNA India. 23 May 2011. Retrieved 27 September 2013.

[9] "RAN 5G Workshop - The Start of Something". 3GPP. September 19, 2015. Retrieved 30 September 2015.

[10] "In 5G proceeding, SpaceX urges FCC to protect future satellite ventures". *FierceWirelessTech*. Retrieved 2015-10-02.

[11] "what is 5g, 5g visions,". *GSM History: History of GSM, Mobile Networks, Vintage Mobiles*.

[12] "Demand Attentive Networks (DAN)".

[13] The Korean IT R&D program of MKE/IITA: 2008-F-004-01 "5G mobile communication systems based on beam-division multiple access and relays with group cooperation".

[14] "5G Innovation Centre". *University of Surrey - Guildford*.

[15] "Mobile communications: Fresh €50 million EU research grants in 2013 to develop '5G' technology". Europa.eu. 26 February 2013. Retrieved 27 September 2013.

[16] "5GrEEn project webpage - Towards Green 5G Mobile Networks". EIT ICT Labs. 15 January 2013. Retrieved 27 September 2013.

[17] Pompili, Dario; Hajisami, Abolfazl; Viswanathan, Hariharasudhan (March 2015). "Dynamic Provisioning and Allocation in Cloud Radio Access Networks (C-RANs)". *Ad Hoc Networks Elsevier* **30**: 128–143.

[18] http://pr.huawei.com/en/news/hw-314871-5g.htm

[19] http://www.timesofmalta.com/articles/view/20150714/local/updated-agreement-for-5g-technology-testing-signed.576618#

[20] http://www.timesofmalta.com/articles/view/20150712/local/pm-thanks-sai-mizzi-as-chinese-telecoms-giant-prepares-to-test-5g-in.576179

[21] "Noord-Groningen krijgt onvoorstelbaar snel mobiel internet". *RTV Noord*. August 2015.

[22] "5GNORMA website".

[23] "mmMAGIC website".

[24] "Xhaul website".

[25] "Flex5Gware website".

[26] J. G. Andrews, S. Buzzi, W. Choi, S. Hanly, A. Lozano, A.C.K. Soong, and J. Zhang, "What will 5G be?," IEEE Journal on Selected Areas in Communications, Vol. 32, No. 6, pp. 1065 - 1082, June 2014.

[27] B. Kouassi, I. Ghauri, L. Deneire, Reciprocity-based cognitive transmissions using a MU massive MIMO approach. IEEE International Conference on Communications (ICC), 2013

[28] T. L. Marzetta (November 2010). "Noncooperative Cellular Wireless with Unlimited Numbers of Base Station Antennas". *IEEE Transactions on Wireless Communications, vol. 9, no. 11*. Bell Labs., Alcatel-Lucent. pp. 56–61, 3590–3600. ISSN 1536-1276. Retrieved 27 September 2013.

[29] J. Hoydis; S. ten Brink; M. Debbah (February 2013). "Massive MIMO in the UL/DL of Cellular Networks: How Many Antennas Do We Need?". *IEEE Journal on Selected Areas in Communications, vol. 31, no. 2*. Bell Labs., Alcatel-Lucent. pp. 160–171. Retrieved 27 September 2013.

[30] Rusek, F.; Persson, D.; Buon Kiong Lau; Larsson, E.G.; Marzetta, T.L.; Edfors, O.; Tufvesson, F. "Scaling Up MIMO: Opportunities and Challenges with Very Large Arrays". *Signal Processing Magazine, IEEE, vol.30, no.1, pp.40,60*. Retrieved Jan 2013.

[31] D. Gesbert; S. Hanly; H. Huang; S. Shamai; O. Simeone; W. Yu (December 2010). "Multi-cell MIMO cooperative networks: A new look at interference". *IEEE Journal on Selected Areas in Communications, vol. 28, no. 9*. EURECOM. pp. 1380–1408. Retrieved 27 September 2013.

[32] Emil Björnson; Eduard Jorswieck (2013). "Optimal Resource Allocation in Coordinated Multi-Cell Systems". *Foundations and Trends in Communications and Information Theory, vol. 9, no. 2-3*. NOW – The Essence of Knowledge. pp. 113–381. Retrieved 27 September 2013.

[33] R. Baldemair; E. Dahlman; G. Fodor; G. Mildh; S. Parkvall; Y. Selen; H. Tullberg; K. Balachandran (March 2013). "Evolving Wireless Communications: Addressing the Challenges and Expectations of the Future". *IEEE Vehicular Technology Magazine, vol. 8, no. 1*. Ericsson Research. pp. 24–30. Retrieved 27 September 2013.

[34] Abdullah Gani; Xichun Li; Lina Yang; Omar Zakaria; Nor Badrul Anuar (February 2009). "Multi-Bandwidth Data Path Design for 5G Wireless Mobile Internets". *WSEAS Transactions on Information Science and Applications archive, Volume 6, Issue 2.* ISSN 1790-0832. Retrieved 27 September 2013.

[35] C. Liang; F. Richard Yu (2014). "Wireless Network Virtualization: A Survey, Some Research Issues and Challenges". *IEEE Communications Surveys & Tutorials.* Retrieved 3 November 2014.

[36] Loretta W. Prencipe (28 February 2003). "Tomorrow's 5g cell phone; Cognitive radio, a 5g device, could forever alter the power balance from wireless service provider to user". *Infoworld Newsletters / Networking.* IDG Group. Retrieved 27 September 2013.

[37] Cornelia-Ionela Badoi; Neeli Prasad; Victor Croitoru; Ramjee Prasad. "5G based cognitive radio". *Wireless Personal Communications, Volume 57, Number 3.* pp. 441–464. doi:10.1007/s11277-010-0082-9. Retrieved 27 September 2013.

[38] Leonardo S. Cardoso; Marco Maso; Mari Kobayashi; Mérouane Debbah (July 2011). "Orthogonal LTE two-tier Cellular Networks" (pdf). *2011 IEEE International Conference on Communications (ICC).* pp. 1–5. Retrieved 27 September 2013.

[39] Toni Janevski (10–13 January 2009). "5G Mobile Phone Concept". *Consumer Communications and Networking Conference, 2009 6th IEEE [1-4244-2308-2].* Facility of Electrical Engineering & Information Technology, University Sv. Kiril i Metodij. Retrieved 27 September 2013.

[40] National Instruments and the University of Edinburgh Collaborate on Massive MIMO Visible Light Communication Networks to Advance 5G, Cambridge Wireless, 20 November 2013

[41] Kelly, Spencer (13 October 2012). "BBC Click Programme - Kenya". BBC News Channel. Retrieved 15 October 2012. Some of the world biggest telecoms firms have joined forces with the UK government to fund a new 5G research center. The facility, to be based at the University of Surrey, will offer testing facilities to operators keen to develop a mobile standard that uses less energy and less radio spectrum, while delivering faster speeds than current 4G technology that's been launched in around 100 countries, including several British cities. They say the new tech could be ready within a decade.

[42] "The University Of Surrey Secures £35M For New 5G Research Centre". University of Surrey. 8 October 2012. Retrieved 15 October 2012.

[43] "5G research centre gets major funding grant". *BBC News* (BBC News Online). 8 October 2012. Retrieved 15 October 2012.

[44] Philipson, Alice (9 October 2012). "Britain aims to join mobile broadband leaders with £35m '5G' research centre". *The Daily Telegraph* (London: Telegraph Media Group). Retrieved 7 January 2013.

[45] "METIS projet presentation" (PDF). November 2012.

[46] "Speech at Mobile World Congress: The Road to 5G". March 2015.

[47] "□□□□, 5□□ □□□□□ □□□□□ □□ □□ □□". 12 May 2013. Retrieved 12 May 2013.

[48] "General METIS presentations available for public".

[49] "India and Israel have agreed to work jointly on development of 5G". *The Times Of India.* 25 July 2013. Retrieved 25 July 2013.

[50] "DoCoMo Wins CEATEC Award for 5G". 3 October 2013. Retrieved 3 October 2013.

[51] Embley, Jochan (6 November 2013). "Huawei plans $600m investment in 10Gbps 5G network". *The Independent* (London). Retrieved 11 November 2013.

[52] "Japan's NTT DoCoMo to Start Testing 5G Mobile Networks". cellular-news. 2014-05-08. Retrieved 2014-05-08.

[53] "Dresden university to inaugurate 5G laboratory". Telecompaper. 25 September 2014. Retrieved 25 September 2014.

[54] ""Мегафон" и Huawei начинают создание сети 5G". CNews. 19 November 2014. Retrieved 19 November 2014.

[55] "Huawei plans to trial 5G mobile internet at the 2018 World Cup". TechRadar. 19 November 2014. Retrieved 19 November 2014.

[56] "SingTel and Huawei Ink MOU to Launch 5G Joint Innoviation Program". Huawei. 19 November 2014. Retrieved 21 November 2014.

[57] "Turkey minister says might cancel 4G tender, switch to 5G: newspaper". Reuters. 28 April 2015. Retrieved 28 April 2015.

[58]  http://www.capital.gr/story/3035583

[59]  "Verizon sets roadmap to 5G technology in U.S.; Field trials to start in 2016". Verizon.  8 September 2015.  Retrieved 9
       September 2015.

[60]  http://www.directmatin.fr/hi-tech/2015-10-01/orange-va-experimenter-la-5g-en-france-712388

## 3.8   External links

- 5G Fifth generation Technology 5G Technology Technical Paper

- Information About Generation 5G

# Chapter 4

# Big data

This article is about large collections of data. For the graph database, see Graph database. For the band, see Big Data (band).

**Big data** is a broad term for data sets so large or complex that traditional data processing applications are inadequate.

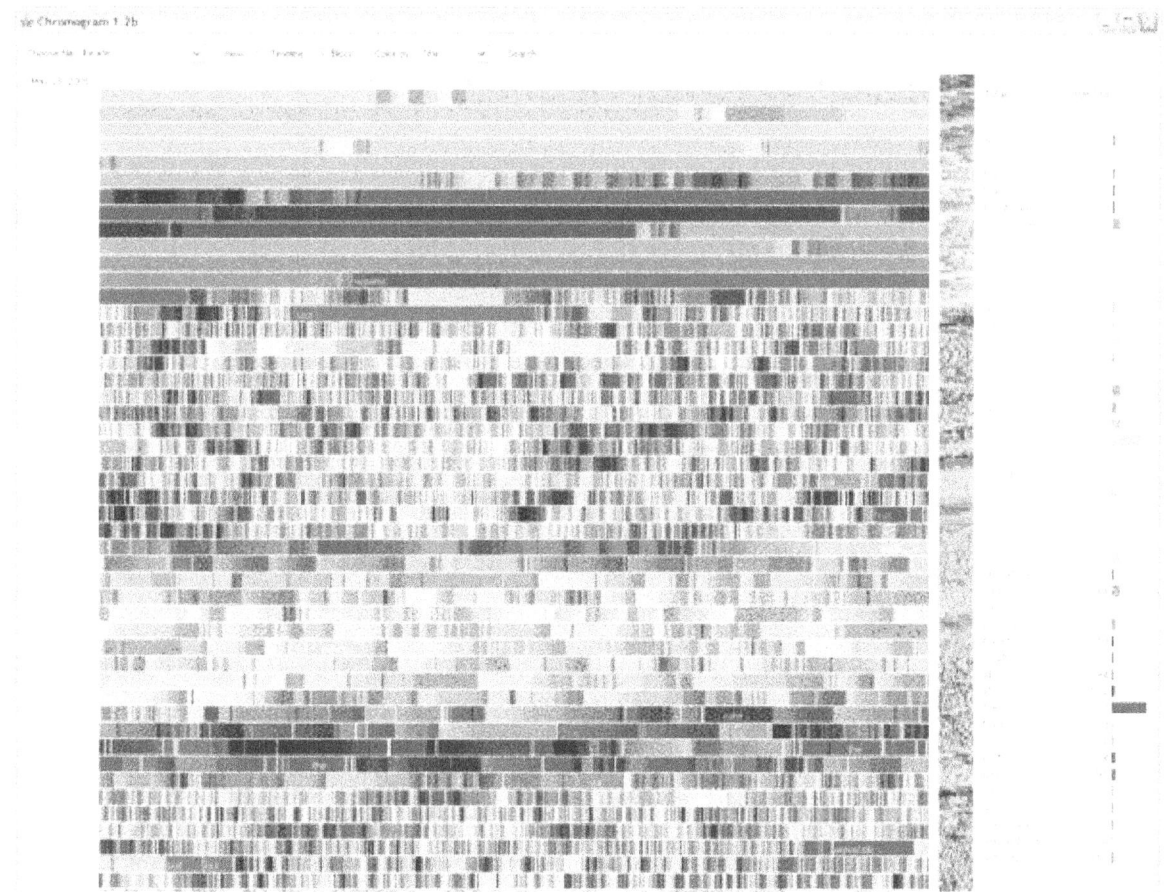

*Visualization of daily Wikipedia edits created by IBM. At multiple terabytes in size, the text and images of Wikipedia are an example of big data.*

Challenges include analysis, capture, data curation, search, sharing, storage, transfer, visualization, and information privacy. The term often refers simply to the use of predictive analytics or other certain advanced methods to extract value

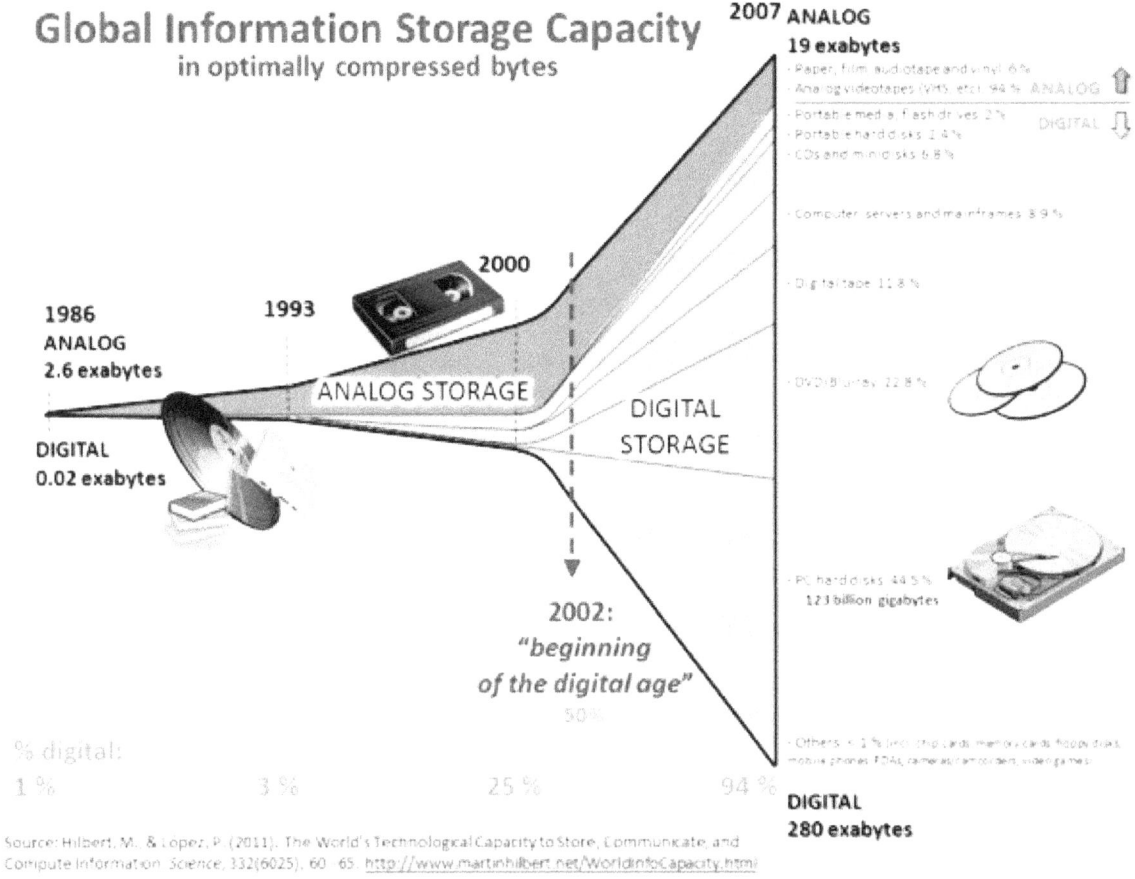

*Growth of and Digitization of Global Information Storage Capacity[11]*

from data, and seldom to a particular size of data set. Accuracy in big data may lead to more confident decision making. And better decisions can mean greater operational efficiency, cost reduction and reduced risk.

Analysis of data sets can find new correlations, to "spot business trends, prevent diseases, combat crime and so on."[2] Scientists, business executives, practitioners of media, and advertising and governments alike regularly meet difficulties with large data sets in areas including Internet search, finance and business informatics. Scientists encounter limitations in e-Science work, including meteorology, genomics,[3] connectomics, complex physics simulations,[4] and biological and environmental research.[5]

Data sets grow in size in part because they are increasingly being gathered by cheap and numerous information-sensing mobile devices, aerial (remote sensing), software logs, cameras, microphones, radio-frequency identification (RFID) readers, and wireless sensor networks.[6][7][8] The world's technological per-capita capacity to store information has roughly doubled every 40 months since the 1980s;[9] as of 2012, every day 2.5 exabytes ($2.5\times10^{18}$) of data were created;[10] The challenge for large enterprises is determining who should own big data initiatives that straddle the entire organization.[11]

Work with big data is necessarily uncommon; most analysis is of "PC size" data, on a desktop PC or notebook[12] that can handle the available data set.

Relational database management systems and desktop statistics and visualization packages often have difficulty handling big data. The work instead requires "massively parallel software running on tens, hundreds, or even thousands of servers".[13] What is considered "big data" varies depending on the capabilities of the users and their tools, and expanding capabilities make big data a moving target. Thus, what is considered "big" one year becomes ordinary later. "For some organizations, facing hundreds of gigabytes of data for the first time may trigger a need to reconsider data management options. For others, it may take tens or hundreds of terabytes before data size becomes a significant consideration."[14]

## 4.1 Definition

Big data usually includes data sets with sizes beyond the ability of commonly used software tools to capture, curate, manage, and process data within a tolerable elapsed time.[15] Big data "size" is a constantly moving target, as of 2012 ranging from a few dozen terabytes to many petabytes of data. Big data is a set of techniques and technologies that require new forms of integration to uncover large hidden values from large datasets that are diverse, complex, and of a massive scale.[16]

In a 2001 research report[17] and related lectures, META Group (now Gartner) analyst Doug Laney defined data growth challenges and opportunities as being three-dimensional, i.e. increasing volume (amount of data), velocity (speed of data in and out), and variety (range of data types and sources). Gartner, and now much of the industry, continue to use this "3Vs" model for describing big data.[18] In 2012, Gartner updated its definition as follows: "Big data is high volume, high velocity, and/or high variety information assets that require new forms of processing to enable enhanced decision making, insight discovery and process optimization."[19] Additionally, a new V "Veracity" is added by some organizations to describe it.[20]

Gartner's definition of the 3Vs is still widely used, and in agreement with a consensual definition that states that "Big Data represents the Information assets characterized by such a High Volume, Velocity and Variety to require specific Technology and Analytical Methods for its transformation into Value".[21] The 3Vs have been expanded to other complementary characteristics of big data:[22][23]

- Volume: big data doesn't sample. It just observes and tracks what happens

- Velocity: big data is often available in real-time

- Variety: big data draws from text, images, audio, video; plus it completes missing pieces through data fusion

- Machine Learning: big data often doesn't ask why and simply detects patterns[24]

- Digital footprint: big data is often a cost-free byproduct of digital interaction[23]

The growing maturity of the concept fosters a more sound difference between big data and Business Intelligence, regarding data and their use:[25]

- Business Intelligence uses descriptive statistics with data with high information density to measure things, detect trends etc.;

- Big data uses inductive statistics and concepts from nonlinear system identification [26] to infer laws (regressions, nonlinear relationships, and causal effects) from large sets of data with low information density[27] to reveal relationships, dependencies and perform predictions of outcomes and behaviors.[26][28]

In a popular tutorial article published in IEEE Access Journal,[29] the authors classified existing definitions of big data into three categories, namely, Attribute Definition, Comparative Definition and Architectural Definition. The authors also presented a big-data technology map that illustrates the key technology evolution for big data.

## 4.2 Characteristics

Big data can be described by the following characteristics:[22][23]

**Volume** The quantity of generated data is important in this context. The size of the data determines the value and potential of the data under consideration, and whether it can actually be considered big data or not. The name 'big data' itself contains a term related to size, and hence the characteristic.

**Variety** The type of content, and an essential fact that data analysts must know. This helps people who are associated with and analyze the data to effectively use the data to their advantage and thus uphold its importance.

**Velocity** In this context, the speed at which the data is generated and processed to meet the demands and the challenges that lie in the path of growth and development.

**Variability** The inconsistency the data can show at times—which can hamper the process of handling and managing the data effectively.

**Veracity** The quality of captured data, which can vary greatly. Accurate analysis depends on the veracity of source data.

**Complexity** Data management can be very complex, especially when large volumes of data come from multiple sources. Data must be linked, connected, and correlated so users can grasp the information the data is supposed to convey.

Factory work and Cyber-physical systems may have a 6C system:

- Connection (sensor and networks)

- Cloud (computing and data on demand)

- Cyber (model and memory)

- Content/context (meaning and correlation)

- Community (sharing and collaboration)

- Customization (personalization and value)

Data must be processed with advanced tools (analytics and algorithms) to reveal meaningful information. Considering visible and invisible issues in, for example, a factory, the information generation algorithm must detect and address invisible issues such as machine degradation, component wear, etc. on the factory floor.[30][31]

## 4.3   Architecture

In 2000, Seisint Inc. developed a C++-based distributed file-sharing framework for data storage and query. The system stores and distributes structured, semi-structured, and unstructured data across multiple servers. Users can build queries in a modified C++ called ECL. ECL uses an "apply schema on read" method to infer the structure of stored data at the time of the query. In 2004, LexisNexis acquired Seisint Inc.[32] and in 2008 acquired ChoicePoint, Inc.[33] and their high-speed parallel processing platform. The two platforms were merged into HPCC Systems and in 2011, HPCC was open-sourced under the Apache v2.0 License. Currently, HPCC and Quantcast File System[34] are the only publicly available platforms capable of analyzing multiple exabytes of data.

In 2004, Google published a paper on a process called MapReduce that used such an architecture. The MapReduce framework provides a parallel processing model and associated implementation to process huge amounts of data. With MapReduce, queries are split and distributed across parallel nodes and processed in parallel (the Map step). The results are then gathered and delivered (the Reduce step). The framework was very successful,[35] so others wanted to replicate the algorithm. Therefore, an implementation of the MapReduce framework was adopted by an Apache open-source project named Hadoop.[36]

MIKE2.0 is an open approach to information management that acknowledges the need for revisions due to big data implications identified in an article titled "Big Data Solution Offering".[37] The methodology addresses handling big data in terms of useful permutations of data sources, complexity in interrelationships, and difficulty in deleting (or modifying) individual records.[38]

Recent studies show that the use of a multiple-layer architecture is an option for dealing with big data. The Distributed Parallel architecture distributes data across multiple processing units, and parallel processing units provide data much faster, by improving processing speeds. This type of architecture inserts data into a parallel DBMS, which implements the use of MapReduce and Hadoop frameworks. This type of framework looks to make the processing power transparent to the end user by using a front-end application server.[39]

Big Data Analytics for Manufacturing Applications can be based on a 5C architecture (connection, conversion, cyber, cognition, and configuration).[40]

The data lake allows an organization to shift its focus from centralized control to a shared model to respond to the changing dynamics of information management. This enables quick segregation of data into the data lake, thereby reducing the overhead time.[41]

## 4.4 Technologies

Big data requires exceptional technologies to efficiently process large quantities of data within tolerable elapsed times. A 2011 McKinsey report[42] suggests suitable technologies include A/B testing, crowdsourcing, data fusion and integration, genetic algorithms, machine learning, natural language processing, signal processing, simulation, time series analysis and visualisation. Multidimensional big data can also be represented as tensors, which can be more efficiently handled by tensor-based computation,[43] such as multilinear subspace learning.[44] Additional technologies being applied to big data include massively parallel-processing (MPP) databases, search-based applications, data mining, distributed file systems, distributed databases, cloud-based infrastructure (applications, storage and computing resources) and the Internet.

Some but not all MPP relational databases have the ability to store and manage petabytes of data. Implicit is the ability to load, monitor, back up, and optimize the use of the large data tables in the RDBMS.[45]

DARPA's Topological Data Analysis program seeks the fundamental structure of massive data sets and in 2008 the technology went public with the launch of a company called Ayasdi.[46]

The practitioners of big data analytics processes are generally hostile to slower shared storage,[47] preferring direct-attached storage (DAS) in its various forms from solid state drive (SSD) to high capacity SATA disk buried inside parallel processing nodes. The perception of shared storage architectures—Storage area network (SAN) and Network-attached storage (NAS) —is that they are relatively slow, complex, and expensive. These qualities are not consistent with big data analytics systems that thrive on system performance, commodity infrastructure, and low cost.

Real or near-real time information delivery is one of the defining characteristics of big data analytics. Latency is therefore avoided whenever and wherever possible. Data in memory is good—data on spinning disk at the other end of a FC SAN connection is not. The cost of a SAN at the scale needed for analytics applications is very much higher than other storage techniques.

There are advantages as well as disadvantages to shared storage in big data analytics, but big data analytics practitioners as of 2011 did not favour it.[48]

## 4.5 Applications

Big data has increased the demand of information management specialists in that Software AG, Oracle Corporation, IBM, Microsoft, SAP, EMC, HP and Dell have spent more than $15 billion on software firms specializing in data management and analytics. In 2010, this industry was worth more than $100 billion and was growing at almost 10 percent a year: about twice as fast as the software business as a whole.[2]

Developed economies increasingly use data-intensive technologies. There are 4.6 billion mobile-phone subscriptions worldwide, and between 1 billion and 2 billion people accessing the internet.[2] Between 1990 and 2005, more than 1 billion people worldwide entered the middle class, which means more people become more literate, which in turn leads to information growth. The world's effective capacity to exchange information through telecommunication networks was 281 petabytes in 1986, 471 petabytes in 1993, 2.2 exabytes in 2000, 65 exabytes in 2007[9] and predictions put the amount of internet traffic at 667 exabytes annually by 2014.[2] According to one estimate, one third of the globally stored information is in the form of alphanumeric text and still image data,[49] which is the format most useful for most big data applications. This also shows the potential of yet unused data (i.e. in the form of video and audio content).

*Bus wrapped with SAP Big data parked outside IDF13.*

While many vendors offer off-the-shelf solutions for Big Data, experts recommend the development of in-house solutions custom-tailored to solve the company's problem at hand if the company has sufficient technical capabilities.[50]

## 4.5.1   Government

The use and adoption of Big Data within governmental processes is beneficial and allows efficiencies in terms of cost, productivity, and innovation. That said, this process does not come without its flaws. Data analysis often requires multiple parts of government (central and local) to work in collaboration and create new and innovative processes to deliver the desired outcome. Below are the thought leading examples within the Governmental Big Data space.

**United States of America**

- In 2012, the Obama administration announced the Big Data Research and Development Initiative, to explore how big data could be used to address important problems faced by the government.[51] The initiative is composed of 84 different big data programs spread across six departments.[52]

- Big data analysis played a large role in Barack Obama's successful 2012 re-election campaign.[53]

- The United States Federal Government owns six of the ten most powerful supercomputers in the world.[54]

- The Utah Data Center is a data center currently being constructed by the United States National Security Agency. When finished, the facility will be able to handle a large amount of information collected by the NSA over the

Internet. The exact amount of storage space is unknown, but more recent sources claim it will be on the order of a few exabytes.[55][56][57]

### India

- Big data analysis helped in parts, responsible for the BJP and its allies to win Indian General Election 2014.[58]

- The Indian Government utilises numerous techniques to ascertain how the Indian electorate is responding to government action, as well as ideas for policy augmentation.

### United Kingdom

Examples of uses of big data in public services:

- Data on prescription drugs: by connecting origin, location and the time of each prescription, a research unit was able to exemplify the considerable delay between the release of any given drug, and a UK-wide adaptation of the National Institute for Health and Care Excellence guidelines. This suggests that new/most up-to-date drugs take some time to filter through to the general patient.

- Joining up data: a local authority blended data about services, such as road gritting rotas, with services for people at risk, such as 'meals on wheels'. The connection of data allowed the local authority to avoid any weather related delay.

## 4.5.2 International development

Research on the effective usage of information and communication technologies for development (also known as ICT4D) suggests that big data technology can make important contributions but also present unique challenges to International development.[59][60] Advancements in big data analysis offer cost-effective opportunities to improve decision-making in critical development areas such as health care, employment, economic productivity, crime, security, and natural disaster and resource management.[61][62][63] However, longstanding challenges for developing regions such as inadequate technological infrastructure and economic and human resource scarcity exacerbate existing concerns with big data such as privacy, imperfect methodology, and interoperability issues.[61]

## 4.5.3 Manufacturing

Based on TCS 2013 Global Trend Study, improvements in supply planning and product quality provide the greatest benefit of big data for manufacturing.[64] Big data provides an infrastructure for transparency in manufacturing industry, which is the ability to unravel uncertainties such as inconsistent component performance and availability. Predictive manufacturing as an applicable approach toward near-zero downtime and transparency requires vast amount of data and advanced prediction tools for a systematic process of data into useful information.[65] A conceptual framework of predictive manufacturing begins with data acquisition where different type of sensory data is available to acquire such as acoustics, vibration, pressure, current, voltage and controller data. Vast amount of sensory data in addition to historical data construct the big data in manufacturing. The generated big data acts as the input into predictive tools and preventive strategies such as Prognostics and Health Management (PHM).

### Cyber-Physical Models

Current PHM implementations mostly utilize data during the actual usage while analytical algorithms can perform more accurately when more information throughout the machine's lifecycle, such as system configuration, physical knowledge and working principles, are included. There is a need to systematically integrate, manage and analyze machinery or process data during different stages of machine life cycle to handle data/information more efficiently and further achieve better transparency of machine health condition for manufacturing industry.

With such motivation a cyber-physical (coupled) model scheme has been developed. The coupled model is a digital twin of the real machine that operates in the cloud platform and simulates the health condition with an integrated knowledge from both data driven analytical algorithms as well as other available physical knowledge. It can also be described as a 5S systematic approach consisting of Sensing, Storage, Synchronization, Synthesis and Service. The coupled model first constructs a digital image from the early design stage. System information and physical knowledge are logged during product design, based on which a simulation model is built as a reference for future analysis. Initial parameters may be statistically generalized and they can be tuned using data from testing or the manufacturing process using parameter estimation. After that step, the simulation model can be considered a mirrored image of the real machine—able to continuously record and track machine condition during the later utilization stage. Finally, with the increased connectivity offered by cloud computing technology, the coupled model also provides better accessibility of machine condition for factory managers in cases where physical access to actual equipment or machine data is limited.[31]

### 4.5.4   Healthcare

Big data analytics has helped healthcare improve by providing personalized medicine and prescriptive analytics, clinical risk intervention and predictive analytics, waste and care variability reduction, automated external and internal reporting of patient data, standardized medical terms and patient registries and fragmented point solutions.

### 4.5.5   Media

**Internet of Things (IoT)**

Main article: Internet of Things

To understand how the media utilises Big Data, it is first necessary to provide some context into the mechanism used for media process. It has been suggested by Nick Couldry and Joseph Turow that practitioners in Media and Advertising approach big data as many actionable points of information about millions of individuals. The industry appears to be moving away from the traditional approach of using specific media environments such as newspapers, magazines, or television shows and instead tap into consumers with technologies that reach targeted people at optimal times in optimal locations. The ultimate aim is to serve, or convey, a message or content that is (statistically speaking) in line with the consumers mindset. For example, publishing environments are increasingly tailoring messages (advertisements) and content (articles) to appeal to consumers that have been exclusively gleaned through various data-mining activities.[66]

- Targeting of consumers (for advertising by marketers)
- Data-capture

Big Data and the IoT work in conjunction. From a media perspective, data is the key derivative of device inter connectivity and allows accurate targeting. The Internet of Things, with the help of big data, therefore transforms the media industry, companies and even governments, opening up a new era of economic growth and competitiveness. The intersection of people, data and intelligent algorithms have far-reaching impacts on media efficiency. The wealth of data generated allows an elaborate layer on the present targeting mechanisms of the industry.

**Technology**

- eBay.com uses two data warehouses at 7.5 petabytes and 40PB as well as a 40PB Hadoop cluster for search, consumer recommendations, and merchandising. Inside eBay's 90PB data warehouse
- Amazon.com handles millions of back-end operations every day, as well as queries from more than half a million third-party sellers. The core technology that keeps Amazon running is Linux-based and as of 2005 they had the world's three largest Linux databases, with capacities of 7.8 TB, 18.5 TB, and 24.7 TB.[67]

- Facebook handles 50 billion photos from its user base.[68]

- As of August 2012, Google was handling roughly 100 billion searches per month.[69]

- Oracle NoSQL Database has been tested to past the 1M ops/sec mark with 8 shards and proceeded to hit 1.2M ops/sec with 10 shards.[70]

### 4.5.6 Private sector

**Retail**

- Walmart handles more than 1 million customer transactions every hour, which are imported into databases estimated to contain more than 2.5 petabytes (2560 terabytes) of data—the equivalent of 167 times the information contained in all the books in the US Library of Congress.[2]

**Retail banking**

- FICO Card Detection System protects accounts world-wide.[71]

- The volume of business data worldwide, across all companies, doubles every 1.2 years, according to estimates.[72][73]

**Real estate**

- Windermere Real Estate uses anonymous GPS signals from nearly 100 million drivers to help new home buyers determine their typical drive times to and from work throughout various times of the day.[74]

### 4.5.7 Science

The Large Hadron Collider experiments represent about 150 million sensors delivering data 40 million times per second. There are nearly 600 million collisions per second. After filtering and refraining from recording more than 99.99995% [75] of these streams, there are 100 collisions of interest per second.[76][77][78]

- As a result, only working with less than 0.001% of the sensor stream data, the data flow from all four LHC experiments represents 25 petabytes annual rate before replication (as of 2012). This becomes nearly 200 petabytes after replication.

- If all sensor data were recorded in LHC, the data flow would be extremely hard to work with. The data flow would exceed 150 million petabytes annual rate, or nearly 500 exabytes per day, before replication. To put the number in perspective, this is equivalent to 500 quintillion ($5\times10^{20}$) bytes per day, almost 200 times more than all the other sources combined in the world.

The Square Kilometre Array is a radio telescope built of thousands of antennas. It is expected to be operational by 2024. Collectively, these antennas are expected to gather 14 exabytes and store one petabyte per day.[79][80] It is considered one of the most ambitious scientific projects ever undertaken.

**Science and research**

- When the Sloan Digital Sky Survey (SDSS) began to collect astronomical data in 2000, it amassed more in its first few weeks than all data collected in the history of astronomy previously. Continuing at a rate of about 200 GB per night, SDSS has amassed more than 140 terabytes of information. When the Large Synoptic Survey Telescope, successor to SDSS, comes online in 2016, its designers expect it to acquire that amount of data every five days.[2]

- Decoding the human genome originally took 10 years to process, now it can be achieved in less than a day. The DNA sequencers have divided the sequencing cost by 10,000 in the last ten years, which is 100 times cheaper than the reduction in cost predicted by Moore's Law.[81]

- The NASA Center for Climate Simulation (NCCS) stores 32 petabytes of climate observations and simulations on the Discover supercomputing cluster.[82]

- Google's DNAStack, compiles and organizes DNA samples of genetic data from around the world to identify diseases and other medical defects. These fast and exact calculations eliminate any 'friction points,' or human errors that could be made by one of the numerous science and biology experts working with the DNA. DNAStack, a part of Google Genomics allows scientists to use the vast sample of resources from Google's search server to scale social experiments that would usually take years, instantly.

## 4.6   Research activities

Encrypted search and cluster formation in big data was demonstrated in March 2014 at the American Society of Engineering Education. Gautam Siwach engaged at *Tackling the challenges of Big Data* by MIT Computer Science and Artificial Intelligence Laboratory and Dr. Amir Esmailpour at UNH Research Group investigated the key features of big data as formation of clusters and their interconnections. They focused on the security of big data and the actual orientation of the term towards the presence of different type of data in an encrypted form at cloud interface by providing the raw definitions and real time examples within the technology. Moreover, they proposed an approach for identifying the encoding technique to advance towards an expedited search over encrypted text leading to the security enhancements in big data.[83]

In March 2012, The White House announced a national "Big Data Initiative" that consisted of six Federal departments and agencies committing more than $200 million to big data research projects.[84]

The initiative included a National Science Foundation "Expeditions in Computing" grant of $10 million over 5 years to the AMPLab[85] at the University of California, Berkeley.[86] The AMPLab also received funds from DARPA, and over a dozen industrial sponsors and uses big data to attack a wide range of problems from predicting traffic congestion[87] to fighting cancer.[88]

The White House Big Data Initiative also included a commitment by the Department of Energy to provide $25 million in funding over 5 years to establish the Scalable Data Management, Analysis and Visualization (SDAV) Institute,[89] led by the Energy Department's Lawrence Berkeley National Laboratory. The SDAV Institute aims to bring together the expertise of six national laboratories and seven universities to develop new tools to help scientists manage and visualize data on the Department's supercomputers.

The U.S. state of Massachusetts announced the Massachusetts Big Data Initiative in May 2012, which provides funding from the state government and private companies to a variety of research institutions.[90] The Massachusetts Institute of Technology hosts the Intel Science and Technology Center for Big Data in the MIT Computer Science and Artificial Intelligence Laboratory, combining government, corporate, and institutional funding and research efforts.[91]

The European Commission is funding the 2-year-long Big Data Public Private Forum through their Seventh Framework Program to engage companies, academics and other stakeholders in discussing big data issues. The project aims to define a strategy in terms of research and innovation to guide supporting actions from the European Commission in the successful implementation of the big data economy. Outcomes of this project will be used as input for Horizon 2020, their next framework program.[92]

The British government announced in March 2014 the founding of the Alan Turing Institute, named after the computer pioneer and code-breaker, which will focus on new ways to collect and analyse large data sets.[93]

At the University of Waterloo Stratford Campus Canadian Open Data Experience (CODE) Inspiration Day, participants demonstrated how using data visualization can increase the understanding and appeal of big data sets and communicate their story to the world.[94]

To make manufacturing more competitive in the United States (and globe), there is a need to integrate more American ingenuity and innovation into manufacturing ; Therefore, National Science Foundation has granted the Industry University cooperative research center for Intelligent Maintenance Systems (IMS) at university of Cincinnati to focus on developing advanced predictive tools and techniques to be applicable in a big data environment.[95] In May 2013, IMS Center held an industry advisory board meeting focusing on big data where presenters from various industrial companies discussed their concerns, issues and future goals in Big Data environment.

Computational social sciences — Anyone can use Application Programming Interfaces (APIs) provided by Big Data holders, such as Google and Twitter, to do research in the social and behavioral sciences.[96] Often these APIs are provided for free.[96] Tobias Preis *et al.* used Google Trends data to demonstrate that Internet users from countries with a higher per capita gross domestic product (GDP) are more likely to search for information about the future than information about the past. The findings suggest there may be a link between online behaviour and real-world economic indicators.[97][98][99] The authors of the study examined Google queries logs made by ratio of the volume of searches for the coming year ('2011') to the volume of searches for the previous year ('2009'), which they call the 'future orientation index'.[100] They compared the future orientation index to the per capita GDP of each country, and found a strong tendency for countries where Google users inquire more about the future to have a higher GDP. The results hint that there may potentially be a relationship between the economic success of a country and the information-seeking behavior of its citizens captured in big data.

Tobias Preis and his colleagues Helen Susannah Moat and H. Eugene Stanley introduced a method to identify online precursors for stock market moves, using trading strategies based on search volume data provided by Google Trends.[101] Their analysis of Google search volume for 98 terms of varying financial relevance, published in *Scientific Reports*,[102] suggests that increases in search volume for financially relevant search terms tend to precede large losses in financial markets.[103][104][105][106][107][108][109][110]

Big data sets come with algorithmic challenges that previously did not exist. Hence, there is a need to fundamentally change the processing ways.[111]

### 4.6.1 Sampling Big Data

An important research question that can be asked about big data sets is whether you need to look at the full data to draw certain conclusions about the properties of the data or is a sample good enough. The name big data itself contains a term related to size and this in an important characteristic of big data. But Sampling (statistics) enables the selection of right data points from within the larger data set to estimate the characteristics of the whole population. For example, there are about 600 million tweets produced every day. Is it necessary to look at all of them to determine the topics that are discussed during the day? Is it necessary to look at all the tweets to determine the sentiment on each of the topics? In manufacturing different types of sensory data such as acoustics, vibration, pressure, current, voltage and controller data are available at short time intervals. To predict down-time it may not be necessary to look at all the data but a sample may be sufficient.

There has been some work done in Sampling algorithms for Big Data. A theoretical formulation for sampling Twitter data has been developed.[112]

## 4.7 Critique

Critiques of the big data paradigm come in two flavors, those that question the implications of the approach itself, and those that question the way it is currently done.

### 4.7.1 Critiques of the big data paradigm

"A crucial problem is that we do not know much about the underlying empirical micro-processes that lead to the emergence of the[se] typical network characteristics of Big Data".[15] In their critique, Snijders, Matzat, and Reips point out that often very strong assumptions are made about mathematical properties that may not at all reflect what is really going on at the level of micro-processes. Mark Graham has leveled broad critiques at Chris Anderson's assertion that big data will spell the end of theory: focusing in particular on the notion that big data must always be contextualized in their social, economic, and political contexts.[113] Even as companies invest eight- and nine-figure sums to derive insight from information streaming in from suppliers and customers, less than 40% of employees have sufficiently mature processes and skills to do so. To overcome this insight deficit, "big data", no matter how comprehensive or well analyzed, must be complemented by "big judgment," according to an article in the Harvard Business Review.[114]

Much in the same line, it has been pointed out that the decisions based on the analysis of big data are inevitably "informed by the world as it was in the past, or, at best, as it currently is".[61] Fed by a large number of data on past experiences, algorithms can predict future development if the future is similar to the past. If the systems dynamics of the future change, the past can say little about the future. For this, it would be necessary to have a thorough understanding of the systems dynamic, which implies theory.[115] As a response to this critique it has been suggested to combine big data approaches with computer simulations, such as agent-based models[61] and Complex Systems. Agent-based models are increasingly getting better in predicting the outcome of social complexities of even unknown future scenarios through computer simulations that are based on a collection of mutually interdependent algorithms.[116][117] In addition, use of multivariate methods that probe for the latent structure of the data, such as factor analysis and cluster analysis, have proven useful as analytic approaches that go well beyond the bi-variate approaches (cross-tabs) typically employed with smaller data sets.

In health and biology, conventional scientific approaches are based on experimentation. For these approaches, the limiting factor is the relevant data that can confirm or refute the initial hypothesis.[118] A new postulate is accepted now in biosciences: the information provided by the data in huge volumes (omics) without prior hypothesis is complementary and sometimes necessary to conventional approaches based on experimentation. In the massive approaches it is the formulation of a relevant hypothesis to explain the data that is the limiting factor. The search logic is reversed and the limits of induction ("Glory of Science and Philosophy scandal", C. D. Broad, 1926) are to be considered.

Privacy advocates are concerned about the threat to privacy represented by increasing storage and integration of personally identifiable information; expert panels have released various policy recommendations to conform practice to expectations of privacy.[119][120][121]

### 4.7.2   Critiques of big data execution

Big data has been called a "fad" in scientific research and its use was even made fun of as an absurd practice in a satirical example on "pig data".[96] Researcher danah boyd has raised concerns about the use of big data in science neglecting principles such as choosing a representative sample by being too concerned about actually handling the huge amounts of data.[122] This approach may lead to results bias in one way or another. Integration across heterogeneous data resources—some that might be considered "big data" and others not—presents formidable logistical as well as analytical challenges, but many researchers argue that such integrations are likely to represent the most promising new frontiers in science.[123] In the provocative article "Critical Questions for Big Data",[124] the authors title big data a part of mythology: "large data sets offer a higher form of intelligence and knowledge [...], with the aura of truth, objectivity, and accuracy". Users of big data are often "lost in the sheer volume of numbers", and "working with Big Data is still subjective, and what it quantifies does not necessarily have a closer claim on objective truth".[124] Recent developments in BI domain, such as pro-active reporting especially target improvements in usability of Big Data, through automated filtering of non-useful data and correlations.[125]

Big data analysis is often shallow compared to analysis of smaller data sets.[126] In many big data projects, there is no large data analysis happening, but the challenge is the extract, transform, load part of data preprocessing.[126]

Big data is a buzzword and a "vague term",[127][128] but at the same time an "obsession"[128] with entrepreneurs, consultants, scientists and the media. Big data showcases such as Google Flu Trends failed to deliver good predictions in recent years, overstating the flu outbreaks by a factor of two. Similarly, Academy awards and election predictions solely based on Twitter were more often off than on target. Big data often poses the same challenges as small data; and adding more data does not solve problems of bias, but may emphasize other problems. In particular data sources such as Twitter are not representative of the overall population, and results drawn from such sources may then lead to wrong conclusions. Google Translate—which is based on big data statistical analysis of text—does a good job at translating web pages. However, results from specialized domains may be dramatically skewed. On the other hand, big data may also introduce new problems, such as the multiple comparisons problem: simultaneously testing a large set of hypotheses is likely to produce many false results that mistakenly appear significant. Ioannidis argued that "most published research findings are false"

[129] due to essentially the same effect: when many scientific teams and researchers each perform many experiments (i.e. process a big amount of scientific data; although not with big data technology), the likelihood of a "significant" result being actually false grows fast – even more so, when only positive results are published.

## 4.8   See also

For a list of companies, and tools, see also: Category:Big data.

- Big Memory
- Data Defined Storage
- Data lineage
- Data science

## 4.9   References

[1] Source

[2] "Data, data everywhere". *The Economist*. 25 February 2010. Retrieved 9 December 2012.

[3] "Community cleverness required". *Nature* **455** (7209): 1. 4 September 2008. doi:10.1038/455001a.

[4] "Sandia sees data management challenges spiral". *HPC Projects*. 4 August 2009.

[5] Reichman, O.J.; Jones, M.B.; Schildhauer, M.P. (2011). "Challenges and Opportunities of Open Data in Ecology". *Science* **331** (6018): 703–5. doi:10.1126/science.1197962. PMID 21311007.

[6] "Data Crush by Christopher Surdak". Retrieved 14 February 2014.

[7] Hellerstein, Joe (9 November 2008). "Parallel Programming in the Age of Big Data". *Gigaom Blog*.

[8] Segaran, Toby; Hammerbacher, Jeff (2009). *Beautiful Data: The Stories Behind Elegant Data Solutions*. O'Reilly Media. p. 257. ISBN 978-0-596-15711-1.

[9] Hilbert & López 2011

[10] "IBM What is big data? — Bringing big data to the enterprise". www.ibm.com. Retrieved 2013-08-26.

[11] Oracle and FSN, "Mastering Big Data: CFO Strategies to Transform Insight into Opportunity", December 2012

[12] "Computing Platforms for Analytics, Data Mining, Data Science". *kdnuggets.com*. Retrieved 15 April 2015.

[13] Jacobs, A. (6 July 2009). "The Pathologies of Big Data". *ACMQueue*.

[14] Magoulas, Roger; Lorica, Ben (February 2009). "Introduction to Big Data". *Release 2.0* (Sebastopol CA: O'Reilly Media) (11).

[15] Snijders, C.; Matzat, U.; Reips, U.-D. (2012). "'Big Data': Big gaps of knowledge in the field of Internet". *International Journal of Internet Science* **7**: 1–5.

[16] Ibrahim; Targio Hashem, Abaker; Yaqoob, Ibrar; Badrul Anuar, Nor; Mokhtar, Salimah; Gani, Abdullah; Ullah Khan, Samee (2015). "big data" on cloud computing: Review and open research issues". *Information Systems* **47**: 98–115. doi:10.1016/j.is.2014.07.006.

[17] Laney, Douglas. "3D Data Management: Controlling Data Volume, Velocity and Variety" (PDF). Gartner. Retrieved 6 February 2001.

[18] Beyer, Mark. "Gartner Says Solving 'Big Data' Challenge Involves More Than Just Managing Volumes of Data". Gartner. Archived from the original on 10 July 2011. Retrieved 13 July 2011.

[19] Laney, Douglas. "The Importance of 'Big Data': A Definition". Gartner. Retrieved 21 June 2012.

[20] "What is Big Data?". Villanova University.

[21] De Mauro, Andrea; Greco, Marco; Grimaldi, Michele (2015). "What is big data? A consensual definition and a review of key research topics". *AIP Conference Proceedings* **1644**: 97–104. doi:10.1063/1.4907823.

[22] Hilbert, Martin. "Big Data for Development: A Review of Promises and Challenges. Development Policy Review.". *martin-hilbert.net*. Retrieved 2015-10-07.

[23] Hilbert, M. (2015). Digital Technology and Social Change [Open Online Course at the University of California] (freely available). https://www.youtube.com/watch?v=XRVIh1h47sA&index=51&list=PLtjBSCvWCU3rNm46D3R85efM0hrzjuAIg Retrieved from https://canvas.instructure.com/courses/949415

[24] Mayer-Schönberger, V., & Cukier, K. (2013). Big data: a revolution that will transform how we live, work and think. London: John Murray.

[25] http://www.bigdataparis.com/presentation/mercredi/PDelort.pdf?PHPSESSID=tv7k70pcr3egpi2r6fi3qbjtj6#page=4

[26] Billings S.A. "Nonlinear System Identification: NARMAX Methods in the Time, Frequency, and Spatio-Temporal Domains". Wiley, 2013

[27] Delort P., Big data Paris 2013 http://www.andsi.fr/tag/dsi-big-data/

[28] Delort P., Big Data car Low-Density Data ? La faible densité en information comme facteur discriminant http://lecercle.lesechos.fr/entrepreneur/tendances-innovation/221169222/big-data-low-density-data-faible-densite-information-com

[29] Hu, Han; Wen, Yonggang; Chua, Tat-Seng; Li, Xuelong (2014). "Towards scalable systems for big data analytics: a technology tutorial". *IEEE Access* **2**: 652–687. doi:10.1109/ACCESS.2014.2332453.

[30] Lee, Jay; Bagheri, Behrad; Kao, Hung-An (2014). "Recent Advances and Trends of Cyber-Physical Systems and Big Data Analytics in Industrial Informatics". *IEEE Int. Conference on Industrial Informatics (INDIN) 2014*.

[31] Lee, Jay; Lapira, Edzel; Bagheri, Behrad; Kao, Hung-an. "Recent advances and trends in predictive manufacturing systems in big data environment". *Manufacturing Letters* **1** (1): 38–41. doi:10.1016/j.mfglet.2013.09.005.

[32] "LexisNexis To Buy Seisint For $775 Million". Washington Post. Retrieved 15 July 2004.

[33] "LexisNexis Parent Set to Buy ChoicePoint". Washington Post. Retrieved 22 February 2008.

[34] "Quantcast Opens Exabyte-Ready File System". www.datanami.com. Retrieved 1 October 2012.

[35] Bertolucci, Jeff "Hadoop: From Experiment To Leading Big Data Platform", "Information Week", 2013. Retrieved on 14 November 2013.

[36] Webster, John. "MapReduce: Simplified Data Processing on Large Clusters", "Search Storage", 2004. Retrieved on 25 March 2013.

[37] "Big Data Solution Offering". MIKE2.0. Retrieved 8 Dec 2013.

[38] "Big Data Definition". MIKE2.0. Retrieved 9 March 2013.

[39] Boja, C; Pocovnicu, A; Bătăgan, L. (2012). "Distributed Parallel Architecture for Big Data". *Informatica Economica* **16** (2): 116–127.

[40] Intelligent Maintenance System

[41] http://www.hcltech.com/sites/default/files/solving_key_businesschallenges_with_big_data_lake_0.pdf

[42] Manyika, James; Chui, Michael; Bughin, Jaques; Brown, Brad; Dobbs, Richard; Roxburgh, Charles; Byers, Angela Hung (May 2011). *Big Data: The next frontier for innovation, competition, and productivity*. McKinsey Global Institute.

[43] "Future Directions in Tensor-Based Computation and Modeling" (PDF). May 2009.

[44] Lu, Haiping; Plataniotis, K.N.; Venetsanopoulos, A.N. (2011). "A Survey of Multilinear Subspace Learning for Tensor Data" (PDF). *Pattern Recognition* **44** (7): 1540–1551. doi:10.1016/j.patcog.2011.01.004.

[45] Monash, Curt (30 April 2009). "eBay's two enormous data warehouses".
Monash, Curt (6 October 2010). "eBay followup — Greenplum out, Teradata > 10 petabytes, Hadoop has some value, and more".

[46] "Resources on how Topological Data Analysis is used to analyze big data". Ayasdi.

[47] CNET News (1 April 2011). "Storage area networks need not apply".

[48] "How New Analytic Systems will Impact Storage". September 2011.

[49] "What Is the Content of the World's Technologically Mediated Information and Communication Capacity: How Much Text, Image, Audio, and Video?", Martin Hilbert (2014), The Information Society; free access to the article through this link: http://www.martinhilbert.net/WhatsTheContent_Hilbert.pdf

[50] Rajpurohit, Anmol (2014-07-11). "Interview: Amy Gershkoff, Director of Customer Analytics & Insights, eBay on How to Design Custom In-House BI Tools". *KDnuggets*. Retrieved 2014-07-14. Dr. Amy Gershkoff: "Generally, I find that off-the-shelf business intelligence tools do not meet the needs of clients who want to derive custom insights from their data. Therefore, for medium-to-large organizations with access to strong technical talent, I usually recommend building custom, in-house solutions."

[51] Kalil, Tom. "Big Data is a Big Deal". White House. Retrieved 26 September 2012.

[52] Executive Office of the President (March 2012). "Big Data Across the Federal Government" (PDF). White House. Retrieved 26 September 2012.

[53] Lampitt, Andrew. "The real story of how big data analytics helped Obama win". *Infoworld*. Retrieved 31 May 2014.

[54] Hoover, J. Nicholas. "Government's 10 Most Powerful Supercomputers". *Information Week*. UBM. Retrieved 26 September 2012.

[55] Bamford, James (15 March 2012). "The NSA Is Building the Country's Biggest Spy Center (Watch What You Say)". *Wired Magazine*. Retrieved 2013-03-18.

[56] "Groundbreaking Ceremony Held for $1.2 Billion Utah Data Center". National Security Agency Central Security Service. Retrieved 2013-03-18.

[57] Hill, Kashmir. "TBlueprints Of NSA's Ridiculously Expensive Data Center In Utah Suggest It Holds Less Info Than Thought". *Forbes*. Retrieved 2013-10-31.

[58] "News: Live Mint". *Are Indian companies making enough sense of Big Data?*. Live Mint – http://www.livemint.com/. 2014-06-23. Retrieved 2014-11-22.

[59] UN GLobal Pulse (2012). Big Data for Development: Opportunities and Challenges (White p. by Letouzé, E.). New York: United Nations

[60] WEF (World Economic Forum), & Vital Wave Consulting. (2012). Big Data, Big Impact: New Possibilities for International Development. World Economic Forum. Retrieved 24 August 2012, from http://www.weforum.org/reports/big-data-big-impact-new-possibilities-ir

[61] "Big Data for Development: From Information- to Knowledge Societies", Martin Hilbert (2013), SSRN Scholarly Paper No. ID 2205145). Rochester, NY: Social Science Research Network; http://papers.ssrn.com/abstract=2205145

[62] "Elena Kvochko, Four Ways To talk About Big Data (Information Communication Technologies for Development Series)". worldbank.org. Retrieved 2012-05-30.

[63] "Daniele Medri: Big Data & Business: An on-going revolution". Statistics Views. 21 Oct 2013.

[64] "Manufacturing: Big Data Benefits and Challenges". *TCS Big Data Study*. Mumbai, India: Tata Consultancy Services Limited. Retrieved 2014-06-03.

[65] Lee, Jay; Wu, F.; Zhao, W.; Ghaffari, M.; Liao, L (Jan 2013). "Prognostics and health management design for rotary machinery systems—Reviews, methodology and applications". *Mechanical Systems and Signal Processing* **42** (1).

[66] Couldry, Nick; Turow, Joseph (2014). "Advertising, Big Data, and the Clearance of the Public Realm: Marketers' New Approaches to the Content Subsidy". *International Journal of Communication* **8**: 1710–1726.

[67] Layton, Julia. "Amazon Technology". Money.howstuffworks.com. Retrieved 2013-03-05.

[68]   "Scaling Facebook to 500 Million Users and Beyond". Facebook.com. Retrieved 2013-07-21.

[69]   "Google Still Doing At Least 1 Trillion Searches Per Year". *Search Engine Land*. 16 January 2015. Retrieved 15 April 2015.

[70]   Lamb, Charles. "Oracle NoSQL Database Exceeds 1 Million Mixed YCSB Ops/Sec".

[71]   "FICO® Falcon® Fraud Manager". Fico.com. Retrieved 2013-07-21.

[72]   "eBay Study: How to Build Trust and Improve the Shopping Experience". Knowwpcarey.com. 2012-05-08. Retrieved 2013-03-05.

[73]   Leading Priorities for Big Data for Business and IT. eMarketer. October 2013. Retrieved January 2014.

[74]   Wingfield, Nick (2013-03-12). "Predicting Commutes More Accurately for Would-Be Home Buyers – NYTimes.com". Bits.blogs.nytimes.cc Retrieved 2013-07-21.

[75]   Alexandru, Dan. "Prof" (PDF). *cds.cern.ch*. CERN. Retrieved 24 March 2015.

[76]   "LHC Brochure, English version. A presentation of the largest and the most powerful particle accelerator in the world, the Large Hadron Collider (LHC), which started up in 2008. Its role, characteristics, technologies, etc. are explained for the general public.". *CERN-Brochure-2010-006-Eng. LHC Brochure, English version*. CERN. Retrieved 20 January 2013.

[77]   "LHC Guide, English version. A collection of facts and figures about the Large Hadron Collider (LHC) in the form of questions and answers.". *CERN-Brochure-2008-001-Eng. LHC Guide, English version*. CERN. Retrieved 20 January 2013.

[78]   Brumfiel, Geoff (19 January 2011). "High-energy physics: Down the petabyte highway". *Nature* **469**. pp. 282–83. doi:10.1038/469282a.

[79]   http://www.zurich.ibm.com/pdf/astron/CeBIT%202013%20Background%20DOME.pdf

[80]   "Future telescope array drives development of exabyte processing". *Ars Technica*. Retrieved 15 April 2015.

[81]   Delort P., OECD ICCP Technology Foresight Forum, 2012. http://www.oecd.org/sti/ieconomy/Session_3_Delort.pdf#page=6

[82]   Webster, Phil. "Supercomputing the Climate: NASA's Big Data Mission". *CSC World*. Computer Sciences Corporation. Retrieved 2013-01-18.

[83]   Siwach, Gautam; Esmailpour, Amir (March 2014). *Encrypted Search & Cluster Formation in Big Data* (PDF). ASEE 2014 Zone I Conference. University of Bridgeport, Bridgeport, Connecticut, USA.

[84]   "Obama Administration Unveils "Big Data" Initiative:Announces $200 Million In New R&D Investments" (PDF). The White House.

[85]   "AMPLab at the University of California, Berkeley". Amplab.cs.berkeley.edu. Retrieved 2013-03-05.

[86]   "NSF Leads Federal Efforts In Big Data". National Science Foundation (NSF). 29 March 2012.

[87]   Timothy Hunter; Teodor Moldovan; Matei Zaharia; Justin Ma; Michael Franklin; Pieter Abbeel; Alexandre Bayen (October 2011). *Scaling the Mobile Millennium System in the Cloud*.

[88]   David Patterson (5 December 2011). "Computer Scientists May Have What It Takes to Help Cure Cancer". The New York Times.

[89]   "Secretary Chu Announces New Institute to Help Scientists Improve Massive Data Set Research on DOE Supercomputers". "energy.gov".

[90]   "Governor Patrick announces new initiative to strengthen Massachusetts' position as a World leader in Big Data". Common-wealth of Massachusetts.

[91]   "Big Data @ CSAIL". Bigdata.csail.mit.edu. 2013-02-22. Retrieved 2013-03-05.

[92]   "Big Data Public Private Forum". Cordis.europa.eu. 2012-09-01. Retrieved 2013-03-05.

[93]   "Alan Turing Institute to be set up to research big data". BBC News. 19 March 2014. Retrieved 2014-03-19.

[94]   "Inspiration day at University of Waterloo, Stratford Campus". http://www.betakit.com/. Retrieved 2014-02-28.

[95] Lee, Jay; Lapira, Edzel; Bagheri, Behrad; Kao, Hung-An (2013). "Recent Advances and Trends in Predictive Manufacturing Systems in Big Data Environment". *Manufacturing Letters* **1** (1): 38–41. doi:10.1016/j.mfglet.2013.09.005.

[96] Reips, Ulf-Dietrich; Matzat, Uwe (2014). "Mining "Big Data" using Big Data Services". *International Journal of Internet Science* **1** (1): 1–8.

[97] Preis, Tobias; Moat,, Helen Susannah; Stanley, H. Eugene; Bishop, Steven R. (2012). "Quantifying the Advantage of Looking Forward". *Scientific Reports* **2**: 350. doi:10.1038/srep00350. PMC 3320057. PMID 22482034.

[98] Marks, Paul (5 April 2012). "Online searches for future linked to economic success". *New Scientist*. Retrieved 9 April 2012.

[99] Johnston, Casey (6 April 2012). "Google Trends reveals clues about the mentality of richer nations". *Ars Technica*. Retrieved 9 April 2012.

[100] Tobias Preis (2012-05-24). "Supplementary Information: The Future Orientation Index is available for download" (PDF). Retrieved 2012-05-24.

[101] Philip Ball (26 April 2013). "Counting Google searches predicts market movements". *Nature*. Retrieved 9 August 2013.

[102] Tobias Preis, Helen Susannah Moat and H. Eugene Stanley (2013). "Quantifying Trading Behavior in Financial Markets Using Google Trends". *Scientific Reports* **3**: 1684. doi:10.1038/srep01684.

[103] Nick Bilton (26 April 2013). "Google Search Terms Can Predict Stock Market, Study Finds". *New York Times*. Retrieved 9 August 2013.

[104] Christopher Matthews (26 April 2013). "Trouble With Your Investment Portfolio? Google It!". *TIME Magazine*. Retrieved 9 August 2013.

[105] Philip Ball (26 April 2013). "Counting Google searches predicts market movements". *Nature*. Retrieved 9 August 2013.

[106] Bernhard Warner (25 April 2013). "'Big Data' Researchers Turn to Google to Beat the Markets". *Bloomberg Businessweek*. Retrieved 9 August 2013.

[107] Hamish McRae (28 April 2013). "Hamish McRae: Need a valuable handle on investor sentiment? Google it". *The Independent* (London). Retrieved 9 August 2013.

[108] Richard Waters (25 April 2013). "Google search proves to be new word in stock market prediction". *Financial Times*. Retrieved 9 August 2013.

[109] David Leinweber (26 April 2013). "Big Data Gets Bigger: Now Google Trends Can Predict The Market". *Forbes*. Retrieved 9 August 2013.

[110] Jason Palmer (25 April 2013). "Google searches predict market moves". *BBC*. Retrieved 9 August 2013.

[111] E. Sejdić, "Adapt current tools for use with big data," *Nature,* vol. vol. 507, no. 7492, pp. 306, Mar. 2014.

[112] Deepan Palguna, Vikas Joshi, Venkatesan Chakaravarthy, Ravi Kothari and L. V. Subramaniam (2015). *Analysis of Sampling Algorithms for Twitter. International Joint Conference on Artificial Intelligence*.

[113] Graham M. (9 March 2012). "Big data and the end of theory?". *The Guardian* (London).

[114] "Good Data Won't Guarantee Good Decisions. Harvard Business Review". *Shah, Shvetank; Horne, Andrew; Capellá, Jaime;*. HBR.org. Retrieved 8 September 2012.

[115] Anderson, C. (2008, 23 June). The End of Theory: The Data Deluge Makes the Scientific Method Obsolete. Wired Magazine, (Science: Discoveries). http://www.wired.com/science/discoveries/magazine/16-07/pb_theory

[116] Rauch, J. (2002). Seeing Around Corners. The Atlantic, (April), 35–48. http://www.theatlantic.com/magazine/archive/2002/04/seeing-around-corners/302471/

[117] Epstein, J. M., & Axtell, R. L. (1996). Growing Artificial Societies: Social Science from the Bottom Up. A Bradford Book.

[118] Delort P., Big data in Biosciences, Big Data Paris, 2012 http://www.bigdataparis.com/documents/Pierre-Delort-INSERM.pdf#page=5

[119] Ohm, Paul. "Don't Build a Database of Ruin". Harvard Business Review.

[120] Darwin Bond-Graham, *Iron Cagebook – The Logical End of Facebook's Patents,* Counterpunch.org, 2013.12.03

[121] Darwin Bond-Graham, *Inside the Tech industry's Startup Conference,* Counterpunch.org, 2013.09.11

[122] danah boyd (2010-04-29). "Privacy and Publicity in the Context of Big Data". *WWW 2010 conference.* Retrieved 2011-04-18.

[123] Jones, MB; Schildhauer, MP; Reichman, OJ; Bowers, S (2006). "The New Bioinformatics: Integrating Ecological Data from the Gene to the Biosphere" (PDF). *Annual Review of Ecology, Evolution, and Systematics* **37** (1): 519–544. doi:10.1146/annurev.ecolsys.37.09130

[124] Boyd, D.; Crawford, K. (2012). "Critical Questions for Big Data". *Information, Communication & Society* **15** (5): 662. doi:10.1080/1369118X.2012.678878.

[125] Failure to Launch: From Big Data to Big Decisions, Forte Wares.

[126] Gregory Piatetsky (2014-08-12). "Interview: Michael Berthold, KNIME Founder, on Research, Creativity, Big Data, and Privacy, Part 2". KDnuggets. Retrieved 2014-08-13.

[127] Pelt, Mason. ""Big Data" is an over used buzzword and this Twitter bot proves it". *siliconangle.com.* SiliconANGLE. Retrieved 4 November 2015.

[128] Harford, Tim (2014-03-28). "Big data: are we making a big mistake?". *Financial Times.* Financial Times. Retrieved 2014-04-07.

[129] Ioannidis, J. P. A. (2005). "Why Most Published Research Findings Are False". *PLoS Medicine* **2** (8): e124. doi:10.1371/journal.pmed.00201 PMC 1182327. PMID 16060722.

## 4.10   Further reading

- Sharma, Sugam; Tim, Udoyara S; Wong, Johnny; Gadia, Shashi; Sharma, Subhash (2014). "A BRIEF REVIEW ON LEADING BIG DATA MODELS". *Data Science Journal* **13**.

- Big Data Computing and Clouds: Challenges, Solutions, and Future Directions. Marcos D. Assuncao, Rodrigo N. Calheiros, Silvia Bianchi, Marco A. S. Netto, Rajkumar Buyya. Technical Report CLOUDS-TR-2013-1, Cloud Computing and Distributed Systems Laboratory, The University of Melbourne, 17 Dec. 2013.

- Encrypted search & cluster formation in Big Data. Gautam Siwach, Dr. A. Esmailpour. American Society for Engineering Education, Conference at the University of Bridgeport, Bridgeport, Connecticut 3–5 April 2014.

- "Big Data for Good" (PDF). ODBMS.org. 5 June 2012. Retrieved 2013-11-12.

- Hilbert, Martin; López, Priscila (2011). "The World's Technological Capacity to Store, Communicate, and Compute Information". *Science* **332** (6025): 60–65. doi:10.1126/science.1200970. PMID 21310967.

- "The Rise of Industrial Big Data". GE Intelligent Platforms. Retrieved 2013-11-12.

- History of Big Data Timeline. A visual history of Big Data with links to supporting articles.

- Hu, Han; Wen, Yonggang; Chua, Tat-Seng; Li, Xuelong (2014). "Towards scalable systems for big data analytics: a technology tutorial". *IEEE Access* **2**: 652–687. doi:10.1109/ACCESS.2014.2332453.

## 4.11   External links

- The dictionary definition of big data at Wiktionary

- MIT Big Data Initiative / Tackling the Challenges of Big Data

# Chapter 5

# Big Memory

**Big Memory** is software and hardware approach that facilitates storing/retrieval/processing of large data sets (terabytes and higher). The term is akin to Big Data and in some instances is a form of Big Data processing architecture implemented in memory rather than in disks/storage. Different Caches are one of the usage of the Big Memory.

The computer memory, namely RAM works orders of magnitude faster than spinning disks or even Solid State Drives. This is usually due to higher raw data throughput because of tighter coupling of CPU and RAM chips (wider bus, CPU and RAM are usually installed on the same motherboard).

Locality of reference is another important characteristic for caches and fast access.

The price of the computer memory chips has significantly declined in the late 2000s. As of 2015 it is affordable to have 256 gigabytes of RAM on a server. [1]

Currently, not many vendors have solid software Big Memory solutions while there are plentiful hardware options (i.e. cheap RAM planks). Terracotta has developed a "in-memory data management suite" [2]

The contemporary software platforms that are based on a garbage collected models, such as .NET CLR , JAVA and others can not usually store hundreds of millions resident objects (object staying in RAM for minutes+ and get promoted to older generations) directly as this results in GC stalls that significantly affect the performance.[3] [4] As of 2015 there are still no efficient simple GC solutions that would have allowed to store 16Gb+ object in a language-native-heaps so there are hybrid approaches emerging for memory-hungry apps in the managed environments.

NFX Unistack framework provides a concept of a large managed memory 'Pile' on a .NET CLR platform based on pre-allocated large byte[] that do not slow GC down.[5] The solution allows to easily store 300,000,000 business objects on a machine with 64 Gb RAM while allowing for millions object put/get transactions a second. The solution was purposely provided for processing Big Memory data sets without going to disk/network.[6] [7] [8] [9]

## 5.1 References

[1] 128 Gb ram chip on NewEgg - http://www.newegg.com/Product/Product.aspx?Item=9SIA7S634M7975

[2] Terracotta, Inc. - http://terracotta.org/products/bigmemorymax

[3] Understanding GC pauses in JVM, HotSpot's minor GC. - http://blog.ragozin.info/2011/06/understanding-gc-pauses-in-jvm-hotspots.html

[4] Long GC pauses in application - http://stackoverflow.com/questions/15696585/long-gc-pauses-in-application

[5] NFX GitHub Inc. - http://github.com/aumcode/nfx

[6] About Managed Object Pile - https://www.youtube.com/watch?v=WFA1XirINB0

[7] .NET Heap with hundreds of millions of objects - https://www.youtube.com/watch?v=Dz_7hukyejQ

[8] Big Memory in .NET 1 - http://www.infoq.com/articles/Big-Memory-Part-1

[9] Big Memory in .NET 2 - http://www.infoq.com/articles/Big-Memory-Part-2

## 5.2   External links

- Media related to Big data at Wikimedia Commons
- The dictionary definition of big data at Wiktionary

# Chapter 6

# Center for Nanotechnology in Society

The **Center for Nanotechnology in Society** at the University of California at Santa Barbara (CNS-UCSB) is funded by the National Science Foundation and "serves as a national research and education center, a network hub among researchers and educators concerned with societal issues concerning nanotechnologies, and a resource base for studying these issues in the US and abroad." The CNS-UCSB began its operations in January 2006.[1][2]

Nanotechnology (sometimes shortened to nanotech or nano) is the study of manipulating matter on an atomic and molecular scale. Generally, nanotechnology deals with structures sized between 1 to 100 nanometre in at least one dimension, and involves developing materials or devices possessing at least one dimension within that size. Quantum mechanical effects are very important at this scale, which is in the quantum realm.[3]

CNS-UCSB looks at the societal implications of nano, including governance, economics, technological development, potential environmental and health risks (risk perception), and "social risks" such as distribution of benefits.[1][2]

## 6.1 History

The Center received its first five years of funding from the U.S. National Science Foundation.[2] The Center aims to disseminate both its technological and social scientific findings on nanoscale science to policymakers and those outside the nano field, and to facilitate broader public participation in the nanotechnological enterprise. It does this through public engagement between academic researchers with regulators, educators, industrial scientists, and policy makers, as well as community-based organizations and NGOs. The Center's education and outreach programs include students and people outside the nanotech field.[1]

## 6.2 Focus

The Center has three main areas of research:[1][2]

1. the historical context of the nano-enterprise;

2. innovation processes and global diffusion of nanotech;

3. risk perception and the public sphere.

## 6.3 Partnerships

CNS–UCSB researchers collaborate with the California NanoSystems Institute, UC Santa Cruz, UC Berkeley, the Chemical Heritage Foundation, Duke University, Rice University, SUNY Levin Institute, and SUNY New Paltz in the US, and

Cardiff University, UK, University of British Columbia, Canada, University of Edinburgh, UK, University of East Anglia, UK, as well as institutes and centers in China and East Asia.[1]

## 6.4   References

[1]  "About CNS–UCSB" Center for Nanotechnology in Society, accessed May 2011.

[2]  Guston, David H. (2010). *Encyclopedia of Nanoscience and Society.* SAGE Publications. pp. 80–82. ISBN 1452266174.

[3]  Cristina Buzea, Ivan Pacheco, and Kevin Robbie (2007). "Nanomaterials and Nanoparticles: Sources and Toxicity". *Biointerphases* **2** (4): MR17–71. doi:10.1116/1.2815690. PMID 20419892.

## 6.5   External links

- Center for Nanotechnology in Society

- CNS Nanoscience and Nanosociety: Risk Innovation Global Energy History

- Innovation Group: Center for Nanotechnology in Society

# Chapter 7

# Center on Nanotechnology and Society

The **Center on Nanotechnology and Society (Nano & Society)** is an affiliate of Illinois Institute of Technology (IIT) and is housed at IIT's Chicago-Kent College of Law. Nano & Society is an affiliate of the Institute on Biotechnology and the Human Future, also based at Chicago-Kent College of Law.

## 7.1 Purpose

The Center is intended for interdisciplinary research, education, and discussion on ethical, legal, economic, policy, and broader social implications of nanoscale science; it approaches nanotechnology with a special focus on the human condition. The Center brings researchers in nano science, law, ethics, medicine, and the social sciences together with leaders in business and industry.

## 7.2 Online offerings

Nano & Society conducts research on NELSI—Nanotechnology's ethical, legal, and social implications. The Center features various initiatives:

**NELSI Global** is a web-based global public policy document archive focused on nanotechnology's ethical, legal and societal implications (NELSI). Located at the Center's website , it serves as a unified clearinghouse, for public policy documents addressing the NELSI agenda. It includes international and U.S.-specific regulations, legislation, case law, congressional testimony, and governmental reports.

## 7.3 Publications, conferences, and events

The Center publishes a monthly e-newsletter, *Nano & Society.* with opinion pieces from guest writers on nanotechnology's potential and potential impact, as well as public attitudes toward and education about nanotechnology; and relevant news updates on nanotechnology.It also publishes *Nanologues*: a print series of booklets on NELSI

The Center hosts the *Chicago Nano Forum,* which encourages public dialogue among nano experts in science, business, social sciences, ethics, and law. These events are webcast on the Center's website.

The *1st Annual Conference on Nanopolicy and the Human Future* provided members of Congress and their staff, researchers, scientists, and others with the latest nanotechnology developments in the ethical, legal and social arena.

## 7.4   External links

- Center on Nanotechnology and Society

- The Institute on Biotechnology and the Human Future

- Chicago-Kent College of Law

- Illinois Institute of Technology

- Nano & Society Resources

Coordinates: 41°52′45″N 87°38′32″W / 41.879102°N 87.64226°W

# Chapter 8

# Clarke's three laws

**Clarke's Three Laws** are three "laws" of prediction formulated by the British science fiction writer Arthur C. Clarke. They are:

1. When a distinguished but elderly scientist states that something is possible, he is almost certainly right. When he states that something is impossible, he is very probably wrong.

2. The only way of discovering the limits of the possible is to venture a little way past them into the impossible.

3. Any sufficiently advanced technology is indistinguishable from magic.

## 8.1   Origins

Clarke's First Law was proposed by Clarke in the essay "Hazards of Prophecy: The Failure of Imagination", in *Profiles of the Future* (1962).[1]

The second law is offered as a simple observation in the same essay. Its status as Clarke's Second Law was conferred by others. In a 1973 revision of *Profiles of the Future*, Clarke acknowledged the Second Law and proposed the Third. "As three laws were good enough for Newton, I have modestly decided to stop there".

The Third Law is the best known and most widely cited, and appears in Clarke's 1973 revision of "Hazards of Prophecy: The Failure of Imagination". It echoes a statement in a 1942 story by Leigh Brackett: "Witchcraft to the ignorant, ... simple science to the learned".[2] An earlier example of this sentiment may be found in Wild Talents (1932) by the author Charles Fort, where he makes the statement: "...a performance that may some day be considered understandable, but that, in these primitive times, so transcends what is said to be the known that it is what I mean by magic." Even earlier, Rider Haggard's novel She (1886) expresses the sentiment multiple times, such as in chapter 17: "Fear not, my Holly, I shall use no magic. Have I not told thee that there is no such thing as magic, though there is such a thing as understanding and applying the forces which are in Nature?"

A fourth law has been added to the canon, despite Clarke's declared intention of not going one better than Newton. Geoff Holder quotes: "For every expert, there is an equal and opposite expert" in his book *101 Things to Do with a Stone Circle* (The History Press, 2009), and offers as his source Clarke's *Profiles of the Future* (new edition, 1999).

## 8.2   Third Law and variants

Clarke's Third Law ("Any sufficiently advanced technology is indistinguishable from magic") has inspired many snowclones and other variations:

- Any sufficiently advanced act of benevolence is indistinguishable from malevolence.[3] (referring to artificial intelligence)

- Any sufficiently advanced incompetence is indistinguishable from malice. (Grey's law)[4] (Hanlon's razor)

- Any sufficiently advanced cluelessness is indistinguishable from malice.[5] (Clark's law)

- Any sufficiently advanced troll is indistinguishable from a genuine kook. (Morgan's maxim)

- Any sufficiently advanced technology is indistinguishable from a rigged demo.[6]

One contrapositive of Clarke's Third Law is

- Any technology distinguishable from magic is insufficiently advanced. (Gehm's corollary)

Clarke's Third Law has been

- reversed for fictional universes involving magic: "Any sufficiently analyzed magic is indistinguishable from science!"[7][8]

- expanded for fictional universes focusing on science fiction: "Any technology, no matter how primitive, is magic to those who don't understand it."[9]

- used to refer to unexplained archaeological finds and reconstructions of folk mysticism: "Any sufficiently ancient recovered wisdom or artifact is also indistinguishable from magic."

## 8.3   See also

- First contact (anthropology)

- Futures studies

- Niven's laws

- Search for extraterrestrial intelligence

- Three Laws of Robotics

- Scientism

## 8.4   References

[1] "'Hazards of Prophecy: The Failure of Imagination'" in the collection *Profiles of the Future: An Enquiry into the Limits of the Possible* (1962, rev. 1973), pp. 14, 21, 36.

[2] "The Sorcerer of Rhiannon", *Astounding* February 1942, p. 39.

[3] Rubin, Charles T. (5 November 2008). "What is the Good of Transhumanism?". In Chadwick, Ruth; Gordijn, Bert. *Medical Enhancement and Posthumanity* (PDF). Springer. p. 149. ISBN 9789048180059. Archived (PDF) from the original on 16 October 2014. Retrieved 17 October 2014.
Rubin is referring to an earlier work of his:
Rubin, Charles T. (1996). "First contact: Copernican moment or nine day's wonder?". In Kingsley, Stuart A.; Lemarchand, Guillermo A. *The Search for Extraterrestrial Intelligence (SETI) in the Optical Spectrum II: 31 January-1 February 1996, San Jose, California, Band 2704*. Proceedings of SPIE - the International Society for Optical Engineering. Bellingham, WA: SPIE—The International Society for Optical Engineering. pp. 161–184. ISBN 9780819420787.

[4] Urban Dictionary: Grey's Law

[5] J. Porter Clark (16 November 1994). "Clark's Law". Newsgroup: alt.news.misc. Retrieved 10 December 2014. They were apologetic and seemed sincere, but sufficiently advanced cluelessness is indistinguishable from malice. 8-)

[6] Quote Details: James Klass: Any sufficiently advanced technology... - The Quotations Page

[7] Girl Genius

[8] Sufficiently Analyzed Magic - TV Tropes

[9] Freefall 00255 November 12, 1999

## 8.5 External links

- The origins of the Three Laws

- "What's Your Law?" (lists some of the corollaries)

- "A Gadget Too Far" at Infinity Plus

# Chapter 9

# Clean Energy Trends

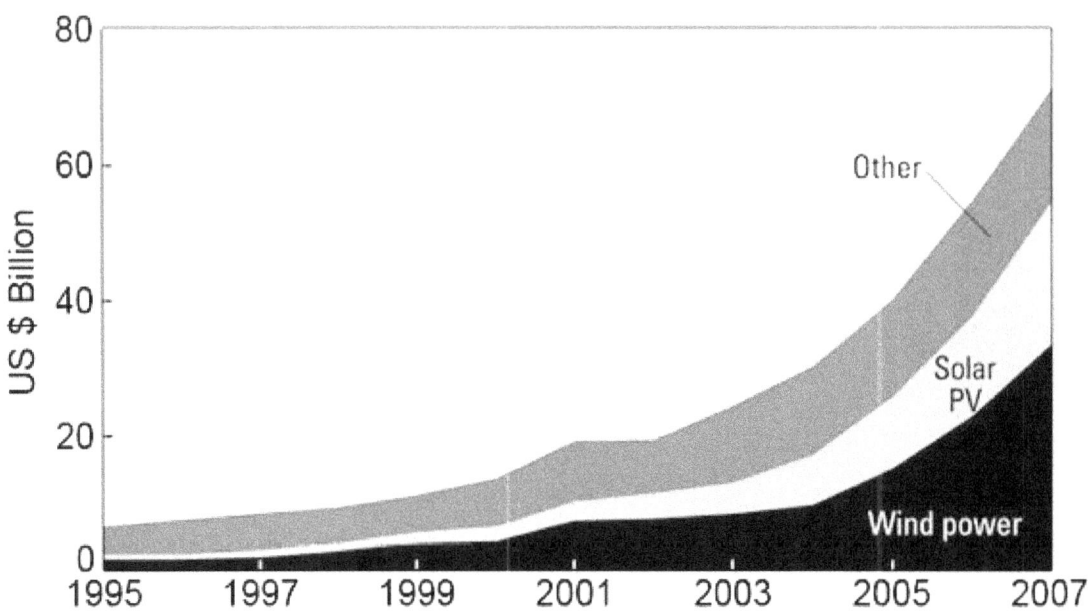

*Global renewable energy investment growth (1995-2007)[1]*

**Clean Energy Trends** is a series of reports by Clean Edge which examine markets for solar, wind, geothermal, fuel cells, biofuels, and other clean energy technologies. Since the publication of the first *Clean Energy Trends* report in 2002, Clean Edge has provided an annual snapshot of both the global and U.S. clean energy sector markets.[2]

## 9.1   2006 trends

In 2006 most climate change skeptics began to change their views. Scientists, investors, business leaders, and politicians moved the agenda from whether climate change was occurring to what should be done about it. The acceptance of climate change as "real" helped to unlock latent interest in clean energy technologies on the part of corporate and political leaders. In Washington and other capitals, clean energy became a bipartisan issue. In corporate boardrooms, it is said to be fast becoming an imperative.[2] And clean energy markets are growing:

"We have reached the point where the steady and rapid growth of clean energy has become an old story. Each year, it seems, brings an ever-higher plateau of success. This appears to be the future of clean energy: a rolling series of technology breakthroughs, landmark corporate investments, industry consolidation, and the not-infrequent emergence of new and sometimes surprising players entering the field."[2]

## 9.2   2007 trends

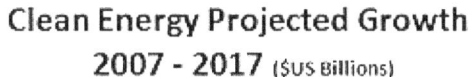

*Projected renewable energy investment growth globally (2007-2017)[3]*

*Clean Energy Trends 2007* shows markets for four benchmark technologies — solar photovoltaics, wind power, biofuels, and fuel cells — continuing their steady climb. Annual revenue for these four technologies increased nearly 39% in one year — to $55 billion in 2006 up from $40 billion in 2005. Clean Edge forecasts that this trajectory will continue to become a $226 billion market by 2016.[4]

Several developments have helped to strengthen clean energy markets in 2007:

- "a near tripling in venture investments in energy technologies in the U.S. to more than $2.4 billion"

- "a new level of commitment by U.S. politicians at the regional, state, and federal levels"

- "significant corporate investments in clean energy acquisitions and expansion initiatives"[2]

## 9.3   See also

- List of energy storage projects

## 9.4   References

[1] REN21 (2008) *Renewables 2007 Global Status Report* (Paris: REN21 Secretariat and Washington, DC:Worldwatch Institute).

[2] Clean Energy Trends 2007

[3]  Makower, J. Pernick, R. Wilder, C. (2008) *Clean Energy Trends 2008* (Clean Edge, USA)

[4]  Renewable Energy Markets to Exceed $220 B by 2016

## 9.5   External links

- Climate for clean energy

- Clean Energy Development Drives Job Creation

- Moving Renewable Energy from the 'Green Ghetto' to Mainstream America

- The Future Ain't What Is Used to Be

# Chapter 10

# Datafication

**Datafication** is a modern technological trend turning many aspects of our life into computerised data [1] and transforming this information into new forms of value. [2] Examples of datafication as applied to social and communication media are how Twitter datafies stray thoughts or datafication of HR by LinkedIn and others. Alternative examples are diverse and include aspects of the built environment, and design via engineering and or other tools that tie data to formal, functional or other physical media outcomes of which Formsolver[3] is an example.

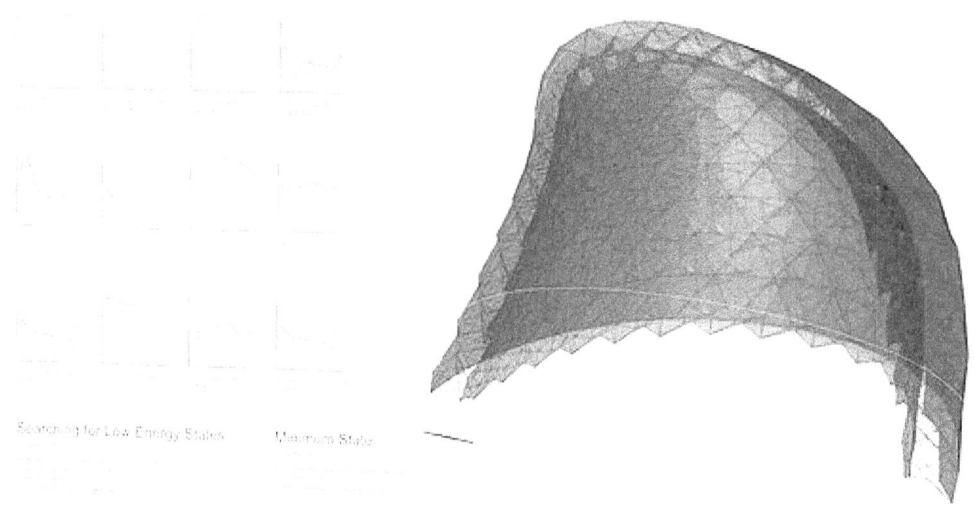

*Example: Datafication of the skin and form of a building to assist engineers, designers and architects determine the performance of particular building geometries. Example provided courtesy of Formsolver.com*

## 10.1   References

[1]  Cukier, Kenneth; Mayer-Schoenberger, Viktor (2013). "The Rise of Big Data". *Foreign Affairs,* (May/June): 28–40. Retrieved 24 January 2014.

[2]  O'Neil, Cathy; Schutt, Rachel (2013). *Doing Data Science.* O'Reilly Media. p. 406. ISBN 978-1-4493-5865-5.

[3]  https://www.formsolver.com

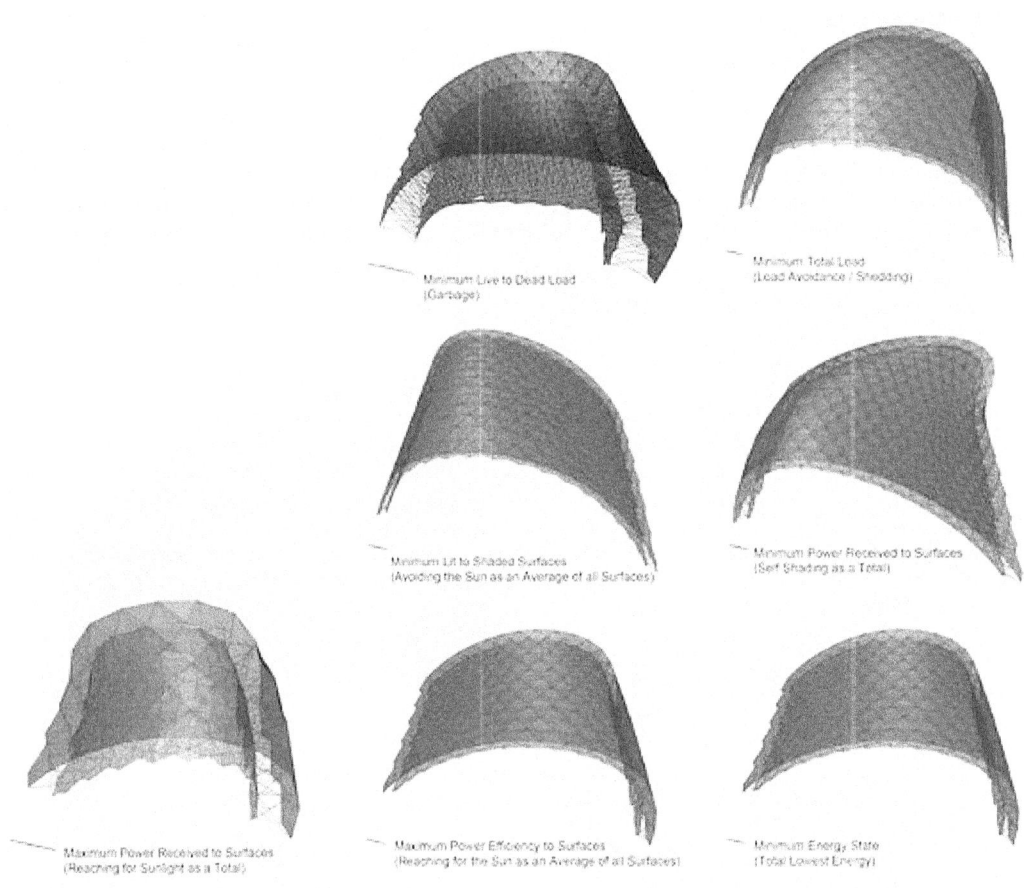

*Example: Shape families resulting from differing goals when data is used for the purposes of shape optimization. Example provided courtesy of Formsolver.com*

# Chapter 11

# Delphi method

The **Delphi method** (/ˈdɛlfaɪ/ *DEL-fy*) is a structured communication technique or method, originally developed as a systematic, interactive forecasting method which relies on a panel of experts.[1][2][3][4] The experts answer questionnaires in two or more rounds. After each round, a facilitator or change agent[5] provides an anonymous summary of the experts' forecasts from the previous round as well as the reasons they provided for their judgments. Thus, experts are encouraged to revise their earlier answers in light of the replies of other members of their panel. It is believed that during this process the range of the answers will decrease and the group will converge towards the "correct" answer. Finally, the process is stopped after a predefined stop criterion (e.g. number of rounds, achievement of consensus, stability of results) and the mean or median scores of the final rounds determine the results.[6]

Delphi is based on the principle that forecasts (or decisions) from a structured group of individuals are more accurate than those from unstructured groups.[7] The technique can also be adapted for use in face-to-face meetings, and is then called mini-Delphi or Estimate-Talk-Estimate (ETE). Delphi has been widely used for business forecasting and has certain advantages over another structured forecasting approach, prediction markets.[8]

## 11.1   History

The name "Delphi" derives from the Oracle of Delphi. The authors of the method were not happy with this name, because it implies "something oracular, something smacking a little of the occult". [9] The Delphi method is based on the assumption that group judgments are more valid than individual judgments.

The Delphi method was developed at the beginning of the Cold War to forecast the impact of technology on warfare.[10] In 1944, General Henry H. Arnold ordered the creation of the report for the U.S. Army Air Corps on the future technological capabilities that might be used by the military.

Different approaches were tried, but the shortcomings of traditional forecasting methods, such as theoretical approach, quantitative models or trend extrapolation, quickly became apparent in areas where precise scientific laws have not been established yet. To combat these shortcomings, the Delphi method was developed by Project RAND during the 1950-1960s (1959) by Olaf Helmer, Norman Dalkey, and Nicholas Rescher.[11] It has been used ever since, together with various modifications and reformulations, such as the Imen-Delphi procedure.

Experts were asked to give their opinion on the probability, frequency, and intensity of possible enemy attacks. Other experts could anonymously give feedback. This process was repeated several times until a consensus emerged.

## 11.2   Key characteristics

The following key characteristics of the Delphi method help the participants to focus on the issues at hand and separate Delphi from other methodologies:

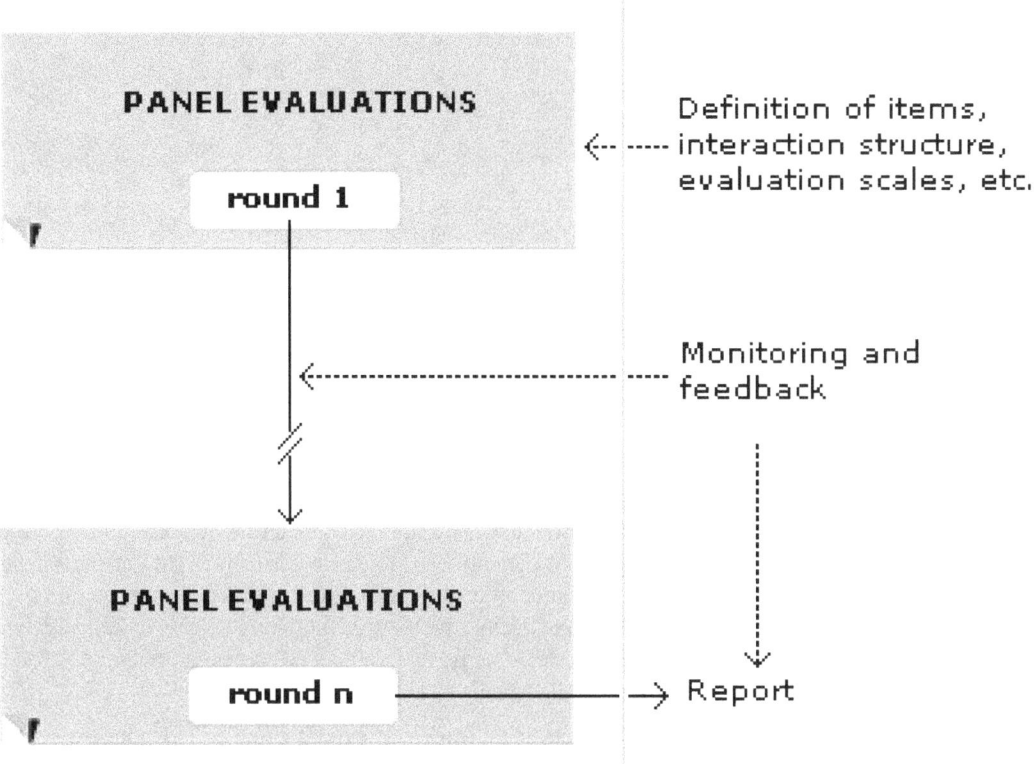

The Delphi Method communication structure

## 11.2.1   Anonymity of the participants

Usually all participants remain anonymous. Their identity is not revealed, even after the completion of the final report. This prevents the authority, personality, or reputation of some participants from dominating others in the process. Arguably, it also frees participants (to some extent) from their personal biases, minimizes the "bandwagon effect" or "halo effect", allows free expression of opinions, encourages open critique, and facilitates admission of errors when revising earlier judgments.

## 11.2.2   Structuring of information flow

The initial contributions from the experts are collected in the form of answers to questionnaires and their comments to these answers. The panel director controls the interactions among the participants by processing the information and filtering out irrelevant content. This avoids the negative effects of face-to-face panel discussions and solves the usual problems of group dynamics.

## 11.2.3   Regular feedback

Participants comment on their own forecasts, the responses of others and on the progress of the panel as a whole. At any moment they can revise their earlier statements. While in regular group meetings participants tend to stick to previously stated opinions and often conform too much to the group leader; the Delphi method prevents it.

### 11.2.4  Role of the facilitator

The person coordinating the Delphi method is usually known as a *facilitator* or Leader, and facilitates the responses of their *panel of experts*, who are selected for a reason, usually that they hold knowledge on an opinion or view. The facilitator sends out questionnaires, surveys etc. and if the panel of experts accept, they follow instructions and present their views. Responses are collected and analyzed, then common and conflicting viewpoints are identified. If consensus is not reached, the process continues through thesis and antithesis, to gradually work towards synthesis, and building consensus.

## 11.3  Applications

### 11.3.1  Use in forecasting

First applications of the Delphi method were in the field of science and technology forecasting. The objective of the method was to combine expert opinions on likelihood and expected development time, of the particular technology, in a single indicator. One of the first such reports, prepared in 1964 by Gordon and Helmer, assessed the direction of long-term trends in science and technology development, covering such topics as scientific breakthroughs, population control, automation, space progress, war prevention and weapon systems. Other forecasts of technology were dealing with vehicle-highway systems, industrial robots, intelligent internet, broadband connections, and technology in education.

Later the Delphi method was applied in other places, especially those related to public policy issues, such as economic trends, health and education. It was also applied successfully and with high accuracy in business forecasting. For example, in one case reported by Basu and Schroeder (1977),[12] the Delphi method predicted the sales of a new product during the first two years with inaccuracy of 3–4% compared with actual sales. Quantitative methods produced errors of 10–15%, and traditional unstructured forecast methods had errors of about 20%.

The Delphi method has also been used as a tool to implement multi-stakeholder approaches for participative policy-making in developing countries. The governments of Latin America and the Caribbean have successfully used the Delphi method as an open-ended public-private sector approach to identify the most urgent challenges for their regional ICT-for-development eLAC Action Plans.[13] As a result, governments have widely acknowledged the value of collective intelligence from civil society, academic and private sector participants of the Delphi, especially in a field of rapid change, such as technology policies.

### 11.3.2  Use in policy-making

From the 1970s, the use of the Delphi technique in public policy-making introduces a number of methodological innovations. In particular:

- the need to examine several types of items (not only *forecasting* items but, typically, *issue* items, *goal* items, and *option* items) leads to introducing different evaluation scales which are not used in the standard Delphi. These often include *desirability*, *feasibility* (technical and political) and *probability*, which the analysts can use to outline different scenarios: the *desired* scenario (from desirability), the *potential* scenario (from feasibility) and the *expected* scenario (from probability);

- the complexity of the issues posed in public policy-making leads to give more importance to the arguments supporting the evaluations of the panelists; so these are often invited to list arguments for and against each option item, and sometimes they are given the possibility to suggest new items to be submitted to the panel;

- for the same reason, the scaling methods, which are used to measure panel evaluations, often include more sophisticated approaches such as multi-dimensional scaling.

Further innovations come from the use of computer-based (and later web-based) Delphi conferences. According to Turoff and Hiltz,[14] in computer-based Delphis:

- the iteration structure used in the paper Delphis, which is divided into three or more discrete rounds, can be replaced by a process of continuous (roundless) interaction, enabling panelists to change their evaluations at any time;

- the statistical group response can be updated in real-time, and shown whenever a panelist provides a new evaluation.

According to Bolognini,[15] web-based Delphis offer two further possibilities, relevant in the context of interactive policy-making and e-democracy.  These are:

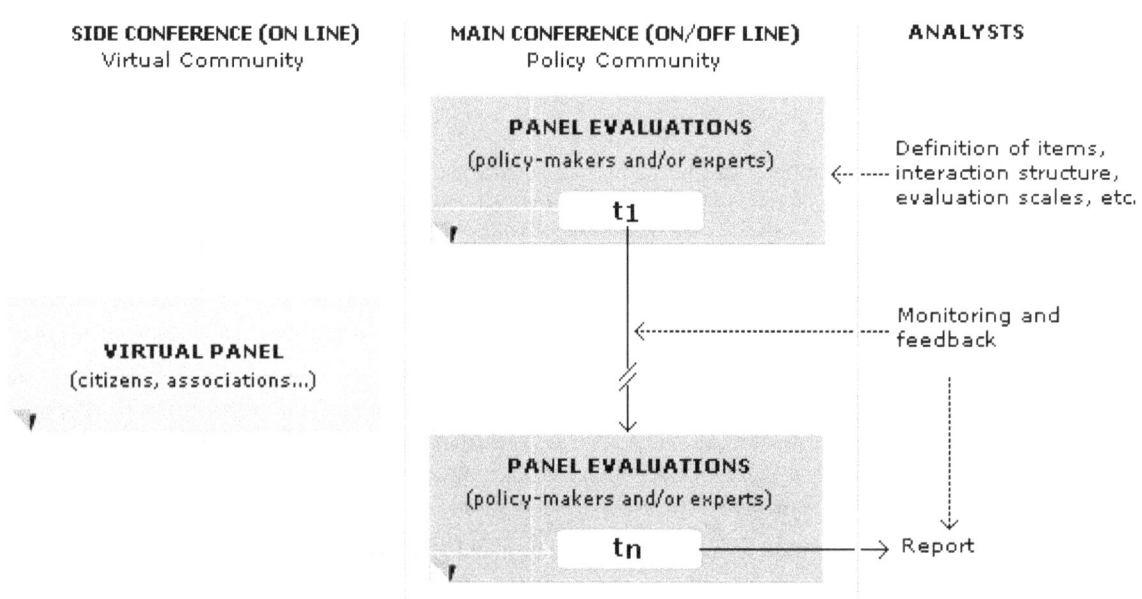

A web-based communication structure (Hyperdelphi).[15]

- the involvement of a large number of participants,

- the use of two or more panels representing different groups (such as policy-makers, experts, citizens), which the administrator can give tasks reflecting their diverse roles and expertise, and make them to interact within ad hoc communication structures.  For example, the *policy community* members (policy-makers and experts) may interact as part of the *main conference* panel, while they receive inputs from a *virtual community* (citizens, associations etc.) involved in a *side conference*.  These web-based variable communication structures, which he calls *Hyperdelphi* (HD), are designed to make Delphi conferences "more fluid and adapted to the hypertextual and interactive nature of digital communication".

One successful example of a (partially) web-based policy Delphi is the five-round Delphi exercise (with 1,454 contributions) for the creation of the eLAC Action Plans in Latin America.  It is believed to be the most extensive online participatory policy-making foresight exercise in the history of intergovernmental processes in the developing world at this time.[13] In addition to the specific policy guidance provided, the authors list the following lessons learned include "(1) the potential of Policy Delphi methods to introduce transparency and accountability into public decision-making, especially in developing countries; (2) the utility of foresight exercises to foster multi-agency networking in the development community; (3) the usefulness of embedding foresight exercises into established mechanisms of representative democracy and international multilateralism, such as the United Nations; (4) the potential of online tools to facilitate participation in resource-scarce developing countries; and (5) the resource-efficiency stemming from the scale of international foresight exercises, and therefore its adequacy for resource-scarce regions."[13]

### 11.3.3 Online Delphi systems

Main article: Real-time Delphi

A number of Delphi forecasts are conducted using web sites that allow the process to be conducted in real-time. For instance, the TechCast Project uses a panel of 100 experts worldwide to forecast breakthroughs in all fields of science and technology. Another example is the Horizon Project, where educational futurists collaborate online using the Delphi method to come up with the technological advancements to look out for in education for the next few years.

## 11.4 Variations

Traditionally the Delphi method has aimed at a consensus of the most probable future by iteration. Other versions, such as the Policy Delphi,[16][17] is instead a decision support method aiming at structuring and discussing the diverse views of the preferred future. In Europe, more recent web-based experiments have used the Delphi method as a communication technique for interactive decision-making and e-democracy.[18] The Argument Delphi, was developed by Osmo Kuusi, focuses on ongoing discussion and finding relevant arguments rather than focusing on the output. The Disaggregative Policy Delphi, developed by Petri Tapio, uses cluster analysis as a systematic tool to construct various scenarios of the future in the latest Delphi round.[19] The respondent's view on the probable and the preferable future are dealt with as separate cases. The computerization of Argument Delphi is relatively difficult because of several problems like argument resolution, argument aggregation and argument evaluation. The computerization of Argument Delphi, was developed by Sadi Evren Seker, proposes solutions to such problems.[20]

## 11.5 Discussion

Overall the track record of the Delphi method is mixed. There have been many cases when the method produced poor results. Still, some authors attribute this to poor application of the method and not to the weaknesses of the method itself. It must also be realized that in areas such as science and technology forecasting, the degree of uncertainty is so great that exact and always correct predictions are impossible, so a high degree of error is to be expected.

Another particular weakness of the Delphi method is that future developments are not always predicted correctly by consensus of experts. The issue of ignorance is important. If panelists are misinformed about a topic, the use of Delphi may only add confidence to their ignorance.[21]

One of the initial problems of the method was its inability to make complex forecasts with multiple factors. Potential future outcomes were usually considered as if they had no effect on each other. Later on, several extensions to the Delphi method were developed to address this problem, such as cross impact analysis, that takes into consideration the possibility that the occurrence of one event may change probabilities of other events covered in the survey. Still the Delphi method can be used most successfully in forecasting single scalar indicators.

Despite these shortcomings, today the Delphi method is a widely accepted forecasting tool and has been used successfully for thousands of studies in areas varying from technology forecasting to drug abuse.[22]

### 11.5.1 Delphi vs. prediction markets

Delphi has characteristics similar to prediction markets as both are structured approaches that aggregate diverse opinions from groups. Yet, there are differences that may be decisive for their relative applicability for different problems.[8]

Some advantages of prediction markets derive from the possibility to provide incentives for participation.

1. They can motivate people to participate over a long period of time and to reveal their true beliefs.

2. They aggregate information automatically and instantly incorporate new information in the forecast.

3. Participants do not have to be selected and recruited manually by a facilitator. They themselves decide whether to participate if they think their private information is not yet incorporated in the forecast.

Delphi seems to have these advantages over prediction markets:

1. Participants reveal their reasoning

2. It is easier to maintain confidentiality

3. Potentially quicker forecasts if experts are readily available.

## 11.6   See also

- *The Wisdom of Crowds*
- Wideband delphi
- DARPA's Policy Analysis Market
- Reference class forecasting
- Planning poker
- Nominal group technique

## 11.7   References

[1] Dalkey, Norman; Helmer, Olaf (1963). "An Experimental Application of the Delphi Method to the use of experts". *Management Science* **9** (3): 458–467. doi:10.1287/mnsc.9.3.458.

[2] Bernice B. Brown (1968). "Delphi Process: A Methodology Used for the Elicitation of Opinions of Experts.": An earlier paper published by RAND (Document No: P-3925, 1968, 15 pages)

[3] Sackman, H. (1974), "Delphi Assessment: Expert Opinion, Forecasting and Group Process", R-1283-PR, April 1974. Brown, Thomas, "An Experiment in Probabilistic Forecasting", R-944-ARPA, 1972

[4] Harold A. Linstone, Murray Turoff (1975), *The Delphi Method: Techniques and Applications*, Reading, Mass.: Addison-Wesley, ISBN 978-0-201-04294-8

[5] Milbrey W. McLaughlin (1990), *The Rand Change Agent Study Revisited: Macro Perspectives and micro Realities* (PDF), Stanford, CA: Stanford University

[6] Rowe and Wright (1999): The Delphi technique as a forecasting tool: issues and analysis. *International Journal of Forecasting*, Volume 15, Issue 4, October 1999.

[7] Rowe and Wright (2001): Expert Opinions in Forecasting. Role of the Delphi Technique. In: Armstrong (Ed.): *Principles of Forecasting: A Handbook of Researchers and Practitioners*, Boston: Kluwer Academic Publishers.

[8] Green, Armstrong, and Graefe (2007): Methods to Elicit Forecasts from Groups: Delphi and Prediction Markets Compared. Forthcoming in *Foresight: The International Journal of Applied Forecasting* (Fall 2007). PDF format

[9] Adler, Michael & Erio Ziglio (1996) *Gazing Into the Oracle: The Delphi Method and Its Application to Social Policy and Public Health*, (Jessica Kingsley Publishers, 1996). ()

[10] "JVTE v15n2: The Modified Delphi Technique - A Rotational Modification," *Journal of Vocational and Technical Education*, Volume 15 Number 2, Spring 1999, web: VT-edu-JVTE-v15n2: of Delphi Technique developed by Olaf Helmer and Norman Dalkey.

[11] Rescher(1998): *Predicting the Future*, (Albany, NY: State University of New York Press, 1998). (, , )

[12] Basu, Shankar; Roger G. Schroeder (May 1977). "Incorporating Judgments in Sales Forecasts: Application of the Delphi Method at American Hoist & Derrick". *Interfaces* **7** (3): 18–27. doi:10.1287/inte.7.3.18.

[13] Hilbert, Martin; Miles, Ian; Othmer, Julia (2009). "Foresight tools for participative policy-making in inter-governmental processes in developing countries: Lessons learned from the eLAC Policy Priorities Delphi" (PDF). *Technological Forecasting and Social Change* **15** (2): 880–896. doi:10.1016/j.techfore.2009.01.001.

[14] Murray Turoff, Starr Roxanne Hiltz, "Computer-based Delphi processes", in Michael Adler, Erio Ziglio (eds.), Gazing Into the Oracle, op. cit.

[15] Maurizio Bolognini (2001), *Democrazia elettronica. Metodo Delphi e politiche pubbliche (Electronic Democracy. Delphi Method and Public Policy-Making)* (in Italian), Rome: Carocci Editore, ISBN 88-430-2035-8. A summary is also in Jerome C. Glenn, Theodore J. Gordon (eds) (2009), *The Millennium Project. Futures Research Methodology*, New York: Amer Council for the United Nations, ISBN 978-0981894119, chap. 23.

[16] Turoff, Murray. "The Design of a Policy Delphi" (PDF). *Technological Forecasting and Social Change* **2** (2): 1970.

[17] Michael Adler, Erio Ziglio (eds.) (1996), *Gazing Into the Oracle: The Delphi Method and Its Application to Social Policy and Public Health*, London: Kingsley Publishers, ISBN 978-1-85302-104-6

[18] Maurizio Bolognini (2001), *Democrazia elettronica. Metodo Delphi e politiche pubbliche (Electronic Democracy. Delphi Method and Public Policy-Making)* (in Italian), Rome: Carocci Editore, ISBN 88-430-2035-8. An example of e-democracy application is DEMOS (Delphi Mediation Online System), whose prototype was presented at the 3rd Worldwide Forum on Electronic Democracy, in 2002.

[19] Tapio, P (2003). "Disaggregative Policy Delphi: Using cluster analysis as a tool for systematic scenario formation". *Technological Forecasting and Social Change* **70** (1): 83–101. doi:10.1016/S0040-1625(01)00177-9.

[20] Seker, S.E. (2015). "Computerized Argument Delphi Technique". *IEEE Access* **3** (2): 368–380. doi:10.1109/ACCESS.2015.2424703.

[21] Green, K. C., Armstrong, J. S., & Graefe, A. (2007). Methods to elicit forecasts from groups: Delphi and prediction markets compared. Foresight: The International Journal of Applied Forecasting, 8, 17–20.

[22] The Delphi Method:Techniques and Applications,Harold A. Linstone and Murray Turoff, Editors © 2002, Murray Turoff and Harold Linstone, TOC III.B.3. The National Drug-Abuse Policy Delphi: Progress Report and Findings to Date, IRENE ANNE JILLSON {http://is.njit.edu/pubs/delphibook/ch3b3.html}

## 11.8 External links

- *The Delphi Method: Techniques and Applications*, edited by Harold A. Linstone and Murray Turoff — a comprehensive book on Delphi method (free download, 11Mb PDF, 618 pages)

- RAND publications on the Delphi Method Downloadable documents from RAND concerning applications of the Delphi Technique.

- *Principles of Forecasting* A free service to support Delphi forecasting and references are available on this site. However source code is not currently available.

- *Institute for Futures Studies and Knowledge Management* Access to several free studies that illustrate the use of the Real-time Delphi method

- Using the Delphi Method for Qualitative, Participatory Action Research in Health Leadership, this article provides a detailed description of the use of modified Delphi for qualitative, participatory action research.

# Chapter 12

# Differential technological development

**Differential technological development** is a strategy proposed by transhumanist philosopher Nick Bostrom in which societies would seek to influence the sequence in which emerging technologies developed. On this approach, societies would strive to retard the development of harmful technologies and their applications, while accelerating the development of beneficial technologies, especially those that offer protection against the harmful ones.[1]

Paul Christiano believes that while accelerating technological progress appears to be one of the best ways to improve human welfare in the next few decades, a faster rate of growth cannot be equally important for the far future because growth must eventually saturate due to physical limits. Hence, from the perspective of the far future, differential technological development appears more crucial.[2]

Inspired by Bostrom's proposal, Luke Muehlhauser and Anna Salamon suggested a more general project of "differential intellectual progress," in which society advances its wisdom, philosophical sophistication, and understanding of risks faster than its technological power.[3][4]

## 12.1 See also

- Existential risk

## 12.2 Notes

[1] Bostrom, Nick (2002). "Existential Risks: Analyzing Human Extinction Scenarios". Retrieved 2006-02-21.

[2] Christiano, Paul (15 Oct 2014). "On Progress and Prosperity". *Effective Altruism Forum*. Retrieved 21 October 2014.

[3] Muehlhauser, Luke; Anna Salamon (2012). "Intelligence Explosion: Evidence and Import" (PDF). pp. 18–19. Retrieved 29 November 2013.

[4] Muehlhauser, Luke (2013). *Facing the Intelligence Explosion*. Machine Intelligence Research Institute. Retrieved 29 November 2013.

# Chapter 13

# Emerging technologies

For specific emerging technologies, see the List of emerging technologies

**Emerging technologies** are technologies that are perceived as capable of changing the status quo. These are technologies characterized by (i) radical novelty, (ii) relatively fast growth, (iii) coherence, (iv) prominent impact, and (v) uncertainty and ambiguity. In other words, an emerging technology can be defined as "a radically novel and relatively fast growing technology characterised by a certain degree of coherence persisting over time and with the potential to exert a considerable impact on the socio-economic domain(s) which is observed in terms of the composition of actors, institutions and patterns of interactions among those, along with the associated knowledge production processes. Its most prominent impact, however, lies in the future and so in the emergence phase is still somewhat uncertain and ambiguous.".[1]

Emerging technologies include a variety of technologies such as educational technology, information technology, nanotechnology, biotechnology, cognitive science, psychotechnology, robotics, and artificial intelligence.[2]

New technological fields may result from the technological convergence of different systems evolving towards similar goals. Convergence brings previously separate technologies such as voice (and telephony features), data (and productivity applications) and video together so that they share resources and interact with each other, creating new efficiencies.

Emerging technologies are those technical innovations which represent progressive developments within a field for competitive advantage;[3] converging technologies represent previously distinct fields which are in some way moving towards stronger inter-connection and similar goals. However, the opinion on the degree of the impact, status and economic viability of several emerging and converging technologies vary.

## 13.1   History of emerging technologies

Main article: History of technology

In the history of technology, emerging technologies[4][5] are contemporary advances and innovation in various fields of technology.

Over centuries, innovative methods and new technologies are developed and opened up. Some of these technologies are due to theoretical research, and others from commercial research and development.

Technological growth includes incremental developments and disruptive technologies. An example of the former was the gradual roll-out of DVD (digital video disc) as a development intended to follow on from the previous optical technology compact disc. By contrast, disruptive technologies are those where a new method replaces the previous technology and makes it redundant, for example, the replacement of horse-drawn carriages by automobiles.

## 13.2    Emerging technology debates

See also: Technology and society

Many writers, including computer scientist Bill Joy,[6] have identified clusters of technologies that they consider critical to humanity's future. Joy warns that the technology could be used by elites for good or evil. They could use it as "good shepherds" for the rest of humanity, or decide everyone else is superfluous and push for mass extinction of those made unnecessary by technology.[7]

Advocates of the benefits of technological change typically see emerging and converging technologies as offering hope for the betterment of the human condition. Cyberphilosophers Alexander Bard and Jan Söderqvist argue in *The Futurica Trilogy* that while Man himself is basically constant throughout human history (genes change very slowly), all relevant change is rather a direct or indirect result of technological innovation (memes change very fast) since new ideas always emanate from technology use and not the other way around.[8] Man should consequently be regarded as history's main constant and technology as its main variable. However, critics of the risks of technological change, and even some advocates such as transhumanist philosopher Nick Bostrom, warn that some of these technologies could pose dangers, perhaps even contribute to the extinction of humanity itself; i.e., some of them could involve existential risks.[9][10]

Much ethical debate centers on issues of distributive justice in allocating access to beneficial forms of technology. Some thinkers, such as environmental ethicist Bill McKibben, oppose the continuing development of advanced technology partly out of fear that its benefits will be distributed unequally in ways that could worsen the plight of the poor.[11] By contrast, inventor Ray Kurzweil is among techno-utopians who believe that emerging and converging technologies could and will eliminate poverty and abolish suffering.[12]

Some analysts such as Martin Ford, author of *The Lights in the Tunnel: Automation, Accelerating Technology and the Economy of the Future*,[13] argue that as information technology advances, robots and other forms of automation will ultimately result in significant unemployment as machines and software begin to match and exceed the capability of workers to perform most routine jobs.

As robotics and artificial intelligence develop further, even many skilled jobs may be threatened. Technologies such as machine learning[14] may ultimately allow computers to do many knowledge-based jobs that require significant education. This may result in substantial unemployment at all skill levels, stagnant or falling wages for most workers, and increased concentration of income and wealth as the owners of capital capture an ever larger fraction of the economy. This in turn could lead to depressed consumer spending and economic growth as the bulk of the population lacks sufficient discretionary income to purchase the products and services produced by the economy.[15]

See also: Technological innovation system, Technological utopianism and Techno-progressivism
See also: Current research in evolutionary biology, Bioconservatism, Bioethics and Biopolitics

## 13.3    Examples

Main article: List of emerging technologies

### 13.3.1    Artificial intelligence

Main articles: Artificial intelligence and Outline of artificial intelligence

*Artificial intelligence* (*AI*) is the intelligence exhibited by machines or software, and the branch of computer science that develops machines and software with human-like intelligence. Major AI researchers and textbooks define the field as "the study and design of intelligent agents", where an intelligent agent is a system that perceives its environment and takes actions that maximize its chances of success. John McCarthy, who coined the term in 1955, defines it as "the science and engineering of making intelligent machines".

The central problems (or goals) of AI research include reasoning, knowledge, planning, learning, natural language processing (communication), perception and the ability to move and manipulate objects. General intelligence (or "strong AI") is still among the field's long-term goals. Currently popular approaches include deep learning, statistical methods, computational intelligence and traditional symbolic AI. There are an enormous number of tools used in AI, including versions of search and mathematical optimization, logic, methods based on probability and economics, and many others.

### 13.3.2 Cancer vaccines

Main article: Cancer vaccine

A *cancer vaccine* is a vaccine that treats existing cancer or prevents the development of cancer in certain high-risk individuals. Vaccines that treat existing cancer are known as *therapeutic* cancer vaccines. There are currently no vaccines able to prevent cancer in general.

On April 14, 2009, Dendreon Corporation announced that their Phase III clinical trial of Provenge, a cancer vaccine designed to treat prostate cancer, had demonstrated an increase in survival. It received U.S. Food and Drug Administration (FDA) approval for use in the treatment of advanced prostate cancer patients on April 29, 2010.[16] The approval of Provenge has stimulated interest in this type of therapy.[17]

### 13.3.3 *In vitro* meat

Main article: In vitro meat

*In vitro meat*, also called *cultured meat*, *cruelty-free meat*, *shmeat*, and *test-tube meat*, is an animal-flesh product that has never been part of a living animal with exception of the fetal calf serum taken from a slaughtered cow. In the 21st century, several research projects have worked on *in vitro* meat in the laboratory.[18] The first in vitro beefburger, created by a Dutch team, was eaten at a demonstration for the press in London in August 2013.[19] There remain difficulties to be overcome before *in vitro* meat becomes commercially available.[20] Cultured meat is prohibitively expensive, but it is expected that the cost could be reduced to compete with that of conventionally obtained meat as technology improves.[21][22] *In vitro* meat is also an ethical issue. Some argue that it is less objectionable than traditionally obtained meat because it doesn't involve killing and reduces the risk of animal cruelty, while others disagree with eating meat that has not developed naturally.

### 13.3.4 Nanotechnology

Main articles: Nanotechnology and Outline of nanotechnology

*Nanotechnology* (sometimes shortened to *nanotech*) is the manipulation of matter on an atomic, molecular, and supramolecular scale. The earliest, widespread description of nanotechnology[23][24] referred to the particular technological goal of precisely manipulating atoms and molecules for fabrication of macroscale products, also now referred to as molecular nanotechnology. A more generalized description of nanotechnology was subsequently established by the National Nanotechnology Initiative, which defines nanotechnology as the manipulation of matter with at least one dimension sized from 1 to 100 nanometers. This definition reflects the fact that quantum mechanical effects are important at this quantum-realm scale, and so the definition shifted from a particular technological goal to a research category inclusive of all types of research and technologies that deal with the special properties of matter that occur below the given size threshold.

### 13.3.5 Robotics

Main articles: Robotics and Outline of robotics

*Robotics* is the branch of technology that deals with the design, construction, operation, and application of robots,[25] as well as computer systems for their control, sensory feedback, and information processing. These technologies deal with automated machines that can take the place of humans in dangerous environments or manufacturing processes, or resemble humans in appearance, behavior, and/or cognition. Many of today's robots are inspired by nature contributing to the field of bio-inspired robotics.

### 13.3.6   Stem cell therapy

Main article: Stem cell therapy

*Stem cell therapy* is an intervention strategy that introduces new adult stem cells into damaged tissue in order to treat disease or injury. Many medical researchers believe that stem cell treatments have the potential to change the face of human disease and alleviate suffering.[26] The ability of stem cells to self-renew and give rise to subsequent generations with variable degrees of differentiation capacities,[27] offers significant potential for generation of tissues that can potentially replace diseased and damaged areas in the body, with minimal risk of rejection and side effects.

## 13.4   Development of emerging technologies

As innovation drives economic growth, and large economic rewards come from new inventions, a great deal of resources (funding and effort) go into the development of emerging technologies. Some of the sources of these resources are described below...

### 13.4.1   Research and development

Research and development is directed towards the advancement of technology in general, and therefore includes development of emerging technologies. *See also List of countries by research and development spending.*

Applied research is a form of systematic inquiry involving the practical application of science. It accesses and uses some part of the research communities' (the academia's) accumulated theories, knowledge, methods, and techniques, for a specific, often state-, business-, or client-driven purpose.

Science policy is the area of public policy which is concerned with the policies that affect the conduct of the science and research enterprise, including the funding of science, often in pursuance of other national policy goals such as technological innovation to promote commercial product development, weapons development, health care and environmental monitoring.

### 13.4.2   DARPA

The Defense Advanced Research Projects Agency (DARPA) is an agency of the U.S. Department of Defense responsible for the development of emerging technologies for use by the military.

DARPA was created in 1958 as the Advanced Research Projects Agency (ARPA) by President Dwight D. Eisenhower. Its purpose was to formulate and execute research and development projects to expand the frontiers of technology and science, with the aim to reach beyond immediate military requirements.

Projects funded by DARPA have provided significant technologies that influenced many non-military fields, such as computer networking and graphical user interfaces in information technology.

### 13.4.3   Technology competitions and awards

There are awards that provide incentive to push the limits of technology (generally synonymous with emerging technologies). Note that while some of these awards reward achievement after-the-fact via analysis of the merits of technological breakthroughs, others provide incentive via competitions for awards offered for goals yet to be achieved.

The Orteig Prize was a $25,000 award offered in 1919 by French hotelier Raymond Orteig for the first nonstop flight between New York City and Paris. In 1927, underdog Charles Lindbergh won the prize in a modified single-engine Ryan aircraft called the Spirit of St. Louis. In total, nine teams spent $400,000 in pursuit of the Orteig Prize.

The XPRIZE series of awards, public competitions designed and managed by the non-profit organization called the X Prize Foundation, are intended to encourage technological development that could benefit mankind. The most high-profile XPRIZE to date was the $10,000,000 Ansari XPRIZE relating to spacecraft development, which was awarded in 2004 for the development of SpaceShipOne.

The Turing Award is an annual prize given by the Association for Computing Machinery (ACM) to "an individual selected for contributions of a technical nature made to the computing community". It is stipulated that "The contributions should be of lasting and major technical importance to the computer field". The Turing Award is generally recognized as the highest distinction in computer science, and in 2014 grew to $1,000,000.

The Millennium Technology Prize is awarded once every two years by Technology Academy Finland, an independent fund established by Finnish industry and the Finnish state in partnership. The first recipient was Tim Berners-Lee, inventor of the World Wide Web.

In 2003, David Gobel seed-funded the Methuselah Mouse Prize (Mprize) to encourage the development of new life extension therapies in mice, which are genetically similar to humans. So far, three Mouse Prizes have been awarded: one for breaking longevity records to Dr. Andrzej Bartke of Southern Illinois University; one for late-onset rejuvenation strategies to Dr. Stephen Spindler of the University of California; and one to Dr. Z. Dave Sharp for his work with the pharmaceutical rapamycin.

## 13.5 Role of science fiction

Science fiction has criticized developing and future technologies, but also inspires innovation and new technology. This topic has been more often discussed in literary and sociological than in scientific forums. Cinema and media theorist Vivian Sobchack examines the dialogue between science fiction films and technological imagination. Technology impacts artists and how they portray their fictionalized subjects, but the fictional world gives back to science by broadening imagination. *How William Shatner Changed the World* is a documentary that gave a number of real-world examples of actualized technological imaginations. While more prevalent in the early years of science fiction with writers like Arthur C. Clarke, new authors still find ways to make currently impossible technologies seem closer to being realized.

## 13.6 See also

- Foresight

- Futures studies

- Institute for Ethics and Emerging Technologies

- Institute on Biotechnology and the Human Future

- Technological change

  - Accelerating change
    - Moore's law
  - Innovation
  - Technological revolution

- Transhumanism

  - Technological singularity

- Upcoming software

## 13.7   Further reading

**General**

- Giersch, H. (1982). Emerging technologies: Consequences for economic growth, structural change, and employment : symposium 1981. Tübingen: Mohr.

- Jones-Garmil, K. (1997). The wired museum: Emerging technology and changing paradigms. Washington, DC: American Association of Museums.

- Kaldis, Byron (2010). "Converging Technologies". Sage Encyclopedia of Nanotechnology and Society, Thousand Oaks: CA, Sage

- Rotolo, D., Hicks, D., Martin, B. R. (2015) What is an emerging technology? Research Policy 44(10): 1827-1843

**Law and policy**

- Branscomb, L. M. (1993). Empowering technology: Implementing a U.S. strategy. Cambridge, Mass: MIT Press.

- Raysman, R., & Raysman, R. (2002). Emerging technologies and the law: Forms and analysis. Commercial law intellectual property series. New York, N.Y.: Law Journal Press.

**Information and learning**

- Hung, D., & Khine, M. S. (2006). Engaged learning with emerging technologies. Dordrecht: Springer.

- Kendall, K. E. (1999). Emerging information technologies: Improving decisions, cooperation, and infrastructure. Thousand Oaks, Calif: Sage Publications.

**Other**

- Cavin, R. K., & Liu, W. (1996). Emerging technologies: Designing low power digital systems. [New York]: Institute of Electrical and Electronics Engineers.

## 13.8   References

**Footnotes**

[1] Rotolo, D., Hicks, D., Martin, B. R. (2015) What is an emerging technology? Research Policy 44(10): 1827-1843. Available here

[2] other examples of developments described as "emerging technologies" can be found here - O'Reilly Emerging Technology Conference 2008 .

[3] International Congress Innovation and Technology XXI: Strategies and Policies Towards the XXI Century, & Soares, O. D. D. (1997). Innovation and technology: Strategies and policies. Dordrecht: Kluwer Academic.

[4] Emerging Technologies: From Hindsight to Foresight. Edited by Edna F. Einsiedel. UBC Press.

[5] Emerging technologies: where is the federal government on the high tech curve? : hearing before the Subcommittee on Government Management, Information, and Technology of the Committee on Government Reform, House of Representatives, One Hundred Sixth Congress, second session, April 24, 2000

[6] See: *Wired Magazine*, "Why the future doesn't need us",

[7] Joy, Bill (2000). "Why the future doesn't need us". Retrieved 2005-11-14.

[8] http://www.amazon.com/The-Futurica-Trilogy-Alexander-Bard/dp/9187173247

[9] Bostrom, Nick (2002). "Existential risks: analyzing human extinction scenarios". Retrieved 2006-02-21.

[10] Warwick, K: "March of the Machines", University of Illinois Press, 2004

[11] McKibben, Bill (2003). *Enough: Staying Human in an Engineered Age*. Times Books. ISBN 0-8050-7096-6.

[12] Kurzweil, Raymond (2005). *The Singularity Is Near: When Humans Transcend Biology*. Viking Adult. ISBN 0-670-03384-7.

[13] Ford, Martin R. (2009), *The Lights in the Tunnel: Automation, Accelerating Technology and the Economy of the Future*, Acculant Publishing, ISBN 978-1448659814. *(e-book available free online.)*

[14] "Machine Learning: A Job Killer?"

[15] "Will Automation Lead to Economic Collapse?"

[16] "Approval Letter - Provenge". Food and Drug Administration. 2010-04-29.

[17] "What Comes After Dendreon's Provenge?". 18 Oct 2010.

[18] Siegelbaum, D.J. (2008-04-23). "In Search of a Test-Tube Hamburger". *Time*. Retrieved 2009-04-30.

[19] World's first lab-grown burger is eaten in London

[20] Building a $325,000 Burger

[21] Temple, James (2009-02-23). "The Future of Food: The No-kill Carnivore". Portfolio.com. Retrieved 2009-08-07.

[22] Preliminary Economics Study of Cultured Meat, eXmoor Pharma Concepts, 2008

[23] Drexler, K. Eric (1986). *Engines of Creation: The Coming Era of Nanotechnology*. Doubleday. ISBN 0-385-19973-2.

[24] Drexler, K. Eric (1992). *Nanosystems: Molecular Machinery, Manufacturing, and Computation*. New York: John Wiley & Sons. ISBN 0-471-57547-X.

[25] "robotics". Oxford Dictionaries. Retrieved 4 February 2011.

[26] Lindvall, O.; Kokaia, Z. (2006). "Stem cells for the treatment of neurological disorders". *Nature* **441** (7097): 1094–1096. doi:10.1038/nature04960. PMID 16810245.

[27] Weissman IL (January 2000). "Stem cells: units of development, units of regeneration, and units in evolution". *Cell* **100** (1): 157–68. doi:10.1016/S0092-8674(00)81692-X. PMID 10647940. as cited in Gurtner GC; Callaghan MJ; Longaker MT (2007). "Progress and potential for regenerative medicine". *Annu. Rev. Med.* **58**: 299–312. doi:10.1146/annurev.med.58.082405.095329. PMID 17076602.

## 13.9 External links

**Websites**

- *Top 10 emerging technologies of 2015*, Scientific American
- Collaborating on Converging Technologies: Education and Practice
- Converging Technologies NSF-sponsored reports
- EU-funded project CONTECS
- EU-funded project KNOWLEDGE NBIC
- EU-funded summerschools on ethics of emerging technologies
- EU High-Level Expert Group on Converging Technologies

- European Parliament Technology Assessment on Converging Technologies report

- ETC Group

- Institute for Ethics and Emerging Technologies

- Institute on Biotechnology and the Human Future

- Converging Technologies Conference 2010 Website

**Videos**

- BBC Horizon Special - Tomorrow's World (2013) - YouTube

- Web 3.0 on Vimeo

# Chapter 14

# The Future of the Mind

***The Future of the Mind: The Scientific Quest to Understand, Enhance, and Empower the Mind*** is a popular science book by futurist and physicist Michio Kaku.[1] The book was initially published on February 25, 2014 by Doubleday.[2]

## 14.1   Overview

The book discusses various possibilities of advanced technology that can alter the brain and mind. Looking into things such as telepathy, telekinesis, consciousness, artificial intelligence, and transhumanism, the book covers a wide range of topics. In it, Kaku proposes a "spacetime theory of consciousness".

## 14.2   Reception

As of March 16, 2014, *The Future of the Mind* made number one on the New York Times Bestseller List.[3]

## 14.3   References

[1]  "Dreaming in Code: Michio Kaku's 'Future of the Mind'". New York Times. Retrieved 23 June 2014.

[2]  "The Future of the Mind: The Scientific Quest to Understand, Enhance, and Empower the Mind, Hardcover, February 25, 2014, by Michio Kaku". amazon.com. Retrieved 2014-11-26.

[3]  "THE FUTURE OF THE MIND, by Michio Kaku. (Doubleday.)". nytimes.com. Retrieved 2014-11-26.

## 14.4   External links

- Michio Kaku's official website

- Booksite

# Chapter 15

# Futures studies

"Future technology" redirects here. For theoretical and upcoming inventions, see Emerging technologies.

"Futurology" redirects here. For Manic Street Preachers album, see Futurology (album).

For the study of the "futures" financial instrument, see futures contract and futures exchange.

**Futures studies** (also called **futurology**) is the study of postulating possible, probable, and preferable futures and the worldviews and myths that underlie them. There is a debate as to whether this discipline is an art or science. In general, it can be considered as a branch of the social sciences and parallel to the field of history. History studies the past, futures studies considers the future. Futures studies (colloquially called "**futures**" by many of the field's practitioners) seeks to understand what is likely to continue and what could plausibly change. Part of the discipline thus seeks a systematic and pattern-based understanding of past and present, and to determine the likelihood of future events and trends.[1] Unlike the physical sciences where a narrower, more specified system is studied, futures studies concerns a much bigger and more complex world system. The methodology and knowledge are much less proven as compared to natural science or even social science like sociology, economics, and political science.

## 15.1 Overview

Futures studies is an interdisciplinary field, studying yesterday's and today's changes, and aggregating and analyzing both lay and professional strategies and opinions with respect to tomorrow. It includes analyzing the sources, patterns, and causes of change and stability in an attempt to develop foresight and to map possible futures. Around the world the field is variously referred to as **futures studies**, **strategic foresight**, **futuristics**, **futures thinking**, **futuring**, and **futurology**. Futures studies and strategic foresight are the academic field's most commonly used terms in the English-speaking world.

**Foresight** was the original term and was first used in this sense by H.G. Wells in 1932.[2] "Futurology" is a term common in encyclopedias, though it is used almost exclusively by nonpractitioners today, at least in the English-speaking world. "Futurology" is defined as the "study of the future."[3] The term was coined by German professor Ossip K. Flechtheim in the mid-1940s, who proposed it as a new branch of knowledge that would include a new science of probability. This term may have fallen from favor in recent decades because modern practitioners stress the importance of alternative and plural futures, rather than one monolithic future, and the limitations of prediction and probability, versus the creation of possible and preferable futures.

Three factors usually distinguish futures studies from the research conducted by other disciplines (although all of these disciplines overlap, to differing degrees). First, futures studies often examines not only possible but also probable, preferable, and "wild card" futures. Second, futures studies typically attempts to gain a holistic or systemic view based on insights from a range of different disciplines. Third, futures studies challenges and unpacks the assumptions behind dominant and contending views of the future. The future thus is not empty but fraught with hidden assumptions. For example, many people expect the collapse of the Earth's ecosystem in the near future, while others believe the current ecosystem will survive indefinitely. A foresight approach would seek to analyze and highlight the assumptions underpinning such views.

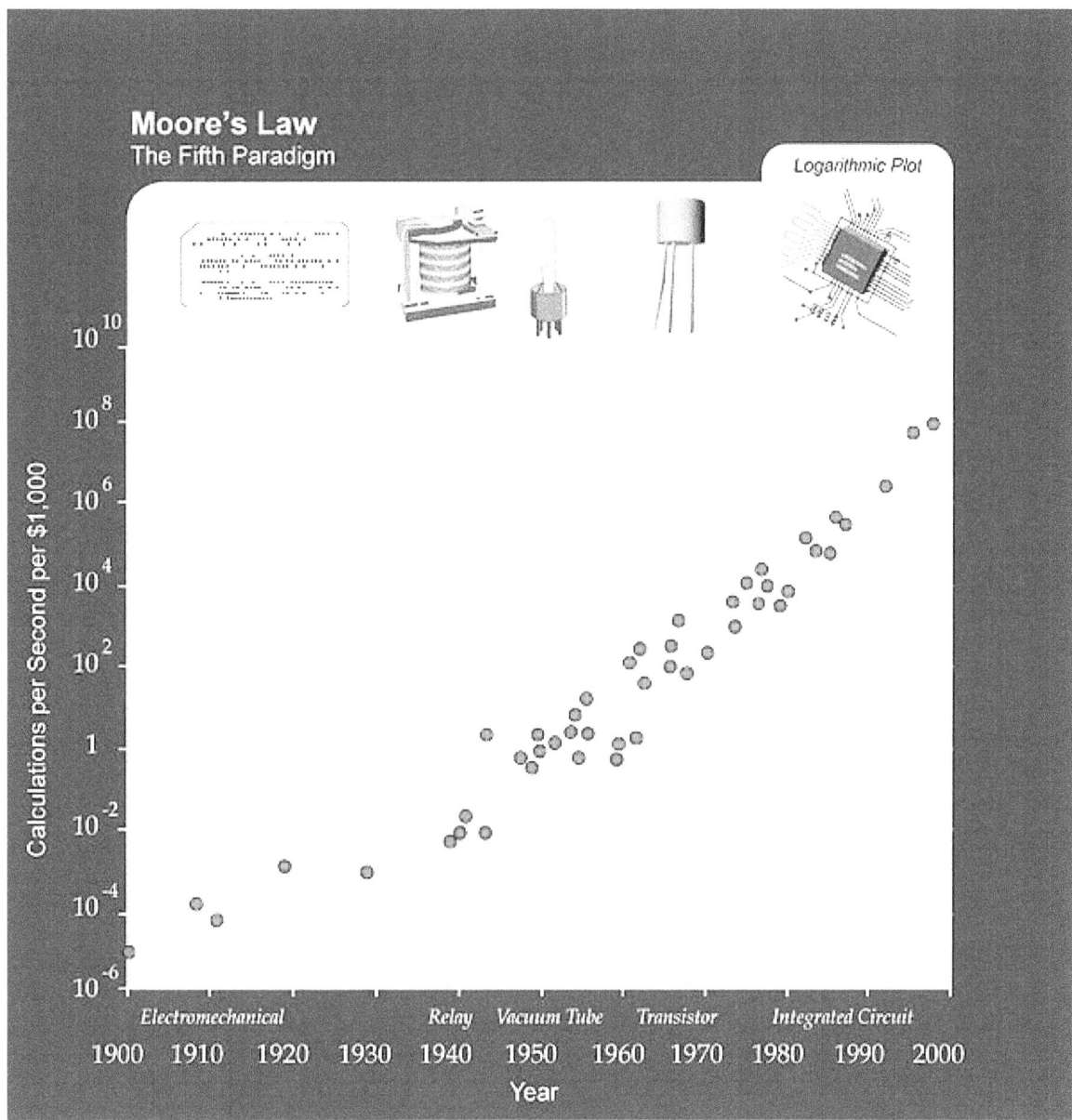

*Moore's law is an example of futures studies; it is a statistical collection of past and present trends with the goal of accurately extrapolating future trends.*

Futures studies does not generally focus on short term predictions such as interest rates over the next business cycle, or of managers or investors with short-term time horizons. Most strategic planning, which develops operational plans for preferred futures with time horizons of one to three years, is also not considered futures. Plans and strategies with longer time horizons that specifically attempt to anticipate possible future events are definitely part of the field.

The futures field also excludes those who make future predictions through professed supernatural means. At the same time, it does seek to understand the models such groups use and the interpretations they give to these models.

## 15.2   History

Johan Galtung and Sohail Inayatullah[4] argue in *Macrohistory and Macrohistorians* that the search for grand patterns of social change goes all the way back to Ssu-Ma Chien (145-90BC) and his theory of the cycles of virtue, although the work of Ibn Khaldun (1332–1406) such as *The Muqaddimah*[5] would be an example that is perhaps more intelligible to modern sociology. Some intellectual foundations of futures studies appeared in the mid-19th century; according to Wendell Bell, Comte's discussion of the metapatterns of social change presages futures studies as a scholarly dialogue.[6]

The first works that attempt to make systematic predictions for the future were written in the 18th century. *Memoirs of the Twentieth Century* written by Samuel Madden in 1733, takes the form of a series of diplomatic letters written in 1997 and 1998 from British representatives in the foreign cities of Constantinople, Rome, Paris, and Moscow.[7] However, the technology of the 20th century is identical to that of Madden's own era - the focus is instead on the political and religious state of the world in the future. Madden went on to write *The Reign of George VI, 1900 to 1925*, where (in the context of the boom in canal construction at the time) he envisioned a large network of waterways that would radically transform patterns of living - "Villages grew into towns and towns became cities".[8]

The genre of science fiction became established towards the end of the 19th century, with notable writers, including Jules Verne and H. G. Wells, setting their stories in an imagined future world.

### 15.2.1   Origins

According to W. Warren Wagar, the founder of future studies was H. G. Wells. His *Anticipations of the Reaction of Mechanical and Scientific Progress Upon Human Life and Thought: An Experiment in Prophecy*, was first serially published in *The Fortnightly Review* in 1901.[9] Anticipating what the world would be like in the year 2000, the book is interesting both for its hits (trains and cars resulting in the dispersion of population from cities to suburbs; moral restrictions declining as men and women seek greater sexual freedom; the defeat of German militarism, the existence of a European Union, and a world order maintained by "English-speaking peoples" based on the urban core between Chicago and New York[10]) and its misses (he did not expect successful aircraft before 1950, and averred that "my imagination refuses to see any sort of submarine doing anything but suffocate its crew and founder at sea").[11][12]

Moving from narrow technological predictions, Wells envisioned the eventual collapse of the capitalist world system after a series of destructive total wars. From this havoc would ultimately emerge a world of peace and plenty, controlled by competent technocrats.[9]

The work was a bestseller, and Wells was invited to deliver a lecture at the Royal Institution in 1902, entitled *The Discovery of the Future*. The lecture was well-received and was soon republished in book form. He advocated for the establishment of a new academic study of the future that would be grounded in scientific methodology rather than just speculation. He argued that a scientifically ordered vision of the future "will be just as certain, just as strictly science, and perhaps just as detailed as the picture that has been built up within the last hundred years to make the geological past." Although conscious of the difficulty in arriving at entirely accurate predictions, he thought that it would still be possible to arrive at a "working knowledge of things in the future".[9]

In his fictional works, Wells predicted the invention and use of the atomic bomb in *The World Set Free* (1914).[13] In *The Shape of Things to Come* (1933) the impending World War and cities destroyed by aerial bombardment was depicted.[14] However, he didn't stop advocating for the establishment of a futures science. In a 1933 BBC broadcast he called for the establishment of "Departments and Professors of Foresight", foreshadowing the development of modern academic futures studies by approximately 40 years.[2]

### 15.2.2   Emergence

Futures studies emerged as an academic discipline in the mid-1960s. First-generation futurists included Herman Kahn, an American Cold War strategist who wrote *On Thermonuclear War* (1960), *Thinking about the unthinkable* (1962) and *The Year 2000: a framework for speculation on the next thirty-three years* (1967); Bertrand de Jouvenel, a French economist who founded Futuribles International in 1960; and Dennis Gabor, a Hungarian-British scientist who wrote *Inventing the Future* (1963) and *The Mature Society. A View of the Future* (1972).[6]

Future studies had a parallel origin with the birth of systems science in academia, and with the idea of national economic and political planning, most notably in France and the Soviet Union.[6][15] In the 1950s, France was continuing to recon-struct their war-torn country. In the process, French scholars, philosophers, writers, and artists searched for what could constitute a more positive future for humanity. The Soviet Union similarly participated in postwar rebuilding, but did so in the context of an established national economic planning process, which also required a long-term, systemic statement of social goals. Future studies was therefore primarily engaged in national planning, and the construction of national symbols.

By contrast, in the United States of America, futures studies as a discipline emerged from the successful application of the tools and perspectives of systems analysis, especially with regard to quartermastering the war-effort. These differing origins account for an initial schism between futures studies in America and futures studies in Europe: U.S. practitioners focused on applied projects, quantitative tools and systems analysis, whereas Europeans preferred to investigate the long-range future of humanity and the Earth, what might constitute that future, what symbols and semantics might express it, and who might articulate these.[16][17]

By the 1960s, academics, philosophers, writers and artists across the globe had begun to explore enough future scenarios so as to fashion a common dialogue. Inventors such as Buckminster Fuller also began highlighting the effect technology might have on global trends as time progressed. This discussion on the intersection of population growth, resource availability and use, economic growth, quality of life, and environmental sustainability – referred to as the "global problematique" – came to wide public attention with the publication of *Limits to Growth*, a study sponsored by the Club of Rome.[18]

### 15.2.3    Further development

International dialogue became institutionalized in the form of the World Futures Studies Federation (WFSF), founded in 1967, with the noted sociologist, Johan Galtung, serving as its first president. In the United States, the publisher Edward Cornish, concerned with these issues, started the World Future Society, an organization focused more on interested laypeople.

1975 saw the founding of the first graduate program in futures studies in the United States, the M.S. program in Futures Studies at the University of Houston–Clear Lake,[19] which moved to the University of Houston in 2007 and renamed the degree to Foresight. There followed a year later the M.A. Program in Public Policy in Alternative Futures at the University of Hawaii at Manoa.[20] The Hawaii program provides particular interest in the light of the schism in perspective between European and U.S. futurists; it bridges that schism by locating futures studies within a pedagogical space defined by neo-Marxism, critical political economic theory, and literary criticism. In the years following the foundation of these two programs, single courses in Futures Studies at all levels of education have proliferated, but complete programs occur only rarely.

As a transdisciplinary field, futures studies attracts generalists. This transdisciplinary nature can also cause problems, owing to it sometimes falling between the cracks of disciplinary boundaries; it also has caused some difficulty in achieving recognition within the traditional curricula of the sciences and the humanities. In contrast to "Futures Studies" at the undergraduate level, some graduate programs in strategic leadership or management offer masters or doctorate programs in "strategic foresight" for mid-career professionals, some even online. Nevertheless, comparatively few new PhDs graduate in Futures Studies each year.

The field currently faces the great challenge of creating a coherent conceptual framework, codified into a well-documented curriculum (or curricula) featuring widely accepted and consistent concepts and theoretical paradigms linked to quantitative and qualitative methods, exemplars of those research methods, and guidelines for their ethical and appropriate application within society. As an indication that previously disparate intellectual dialogues have in fact started converging into a recognizable discipline,[21] at least six solidly-researched and well-accepted first attempts to synthesize a coherent framework for the field have appeared: Eleonora Masini's Why Futures Studies,[22] James Dator's Advancing Futures Studies,[23] Ziauddin Sardar's Rescuing all of our Futures,[24] Sohail Inayatullah's Questioning the future,[25] Richard A. Slaughter's *The Knowledge Base of Futures Studies*,[26] a collection of essays by senior practitioners, and Wendell Bell's two-volume work, *The Foundations of Futures Studies*.[27]

## 15.3    Probability and predictability

Some aspects of the future, such as celestial mechanics, are highly predictable, and may even be described by relatively simple mathematical models. At present however, science has yielded only a special minority of such "easy to predict" physical processes. Theories such as chaos theory, nonlinear science and standard evolutionary theory have allowed us to understand many complex systems as contingent (sensitively dependent on complex environmental conditions) and stochastic (random within constraints), making the vast majority of future events unpredictable, *in any specific case.*

Not surprisingly, the tension between predictability and unpredictability is a source of controversy and conflict among futures studies scholars and practitioners. Some argue that the future is essentially unpredictable, and that "the best way to predict the future is to create it." Others believe, as Flechtheim, that advances in science, probability, modeling and statistics will allow us to continue to improve our understanding of probable futures, while this area presently remains less well developed than methods for exploring possible and preferable futures.

As an example, consider the process of electing the president of the United States. At one level we observe that any U.S. citizen over 35 may run for president, so this process may appear too unconstrained for useful prediction. Yet further investigation demonstrates that only certain public individuals (current and former presidents and vice presidents, senators, state governors, popular military commanders, mayors of very large cities, etc.) receive the appropriate "social credentials" that are historical prerequisites for election. Thus with a minimum of effort at formulating the problem for statistical prediction, a much reduced pool of candidates can be described, improving our probabilistic foresight. Applying further statistical intelligence to this problem, we can observe that in certain election prediction markets such as the Iowa Electronic Markets, reliable forecasts have been generated over long spans of time and conditions, with results superior to individual experts or polls. Such markets, which may be operated publicly or as an internal market, are just one of several promising frontiers in predictive futures research.

Such improvements in the predictability of individual events do not though, from a complexity theory viewpoint, address the unpredictability inherent in dealing with entire systems, which emerge from the interaction between multiple individual events.

## 15.4   Methodologies

Futures practitioners use a wide range of models and methods (theory and practice), many of which come from other academic disciplines, including economics, sociology, geography, history, engineering, mathematics, psychology, technology, tourism, physics, biology, astronomy, and aspects of theology (specifically, the range of future beliefs).

One of the fundamental assumptions in futures studies is that the future is plural not singular, that is, that it consists of alternative futures of varying likelihood but that it is impossible in principle to say with certainty which one will occur. The primary effort in futures studies, therefore, is to identify and describe alternative futures. This effort includes collecting quantitative and qualitative data about the possibility, probability, and desirability of change. The plurality of the term "futures" in futures studies denotes the rich variety of alternative futures, including the subset of preferable futures (normative futures), that can be studied.

Practitioners of the discipline previously concentrated on extrapolating present technological, economic or social trends, or on attempting to predict future trends, but more recently they have started to examine social systems and uncertainties and to build scenarios, question the worldviews behind such scenarios via the causal layered analysis method (and others), create preferred visions of the future, and use backcasting to derive alternative implementation strategies. Apart from extrapolation and scenarios, many dozens of methods and techniques are used in futures research (see below).

Futures studies also includes normative or preferred futures, but a major contribution involves connecting both extrapolated (exploratory) and normative research to help individuals and organisations to build better social futures amid a (presumed) landscape of shifting social changes. Practitioners use varying proportions of inspiration and research. Futures studies only rarely uses the scientific method in the sense of controlled, repeatable and falsifiable experiments with highly standardized methodologies, given that environmental conditions for repeating a predictive scheme are usually quite hard to control. However, many futurists are informed by scientific techniques. Some historians project patterns observed in past civilizations upon present-day society to anticipate what will happen in the future. Oswald Spengler's "Decline of the West" argued, for instance, that western society, like imperial Rome, had reached a stage of cultural maturity that would inexorably lead to decline, in measurable ways.

Futures studies is often summarized as being concerned with "three Ps and a W", or possible, probable, and preferable futures, plus wildcards, which are low probability but high impact events (positive or negative), should they occur. Many futurists, however, do not use the wild card approach. Rather, they use a methodology called Emerging Issues Analysis.

It searches for the seeds of change, issues that are likely to move from unknown to the known, from low impact to high impact.

Estimates of probability are involved with two of the four central concerns of foresight professionals (discerning and classifying both probable and wildcard events), while considering the range of possible futures, recognizing the plurality of existing alternative futures, characterizing and attempting to resolve normative disagreements on the future, and envisioning and creating preferred futures are other major areas of scholarship. Most estimates of probability in futures studies are normative and qualitative, though significant progress on statistical and quantitative methods (technology and information growth curves, cliometrics, predictive psychology, prediction markets, etc.) has been made in recent decades.

## 15.4.1 Futures techniques

Main article: Futures techniques

While forecasting – i.e., attempts to predict future states from current trends – is a common methodology, professional scenarios often rely on "backcasting": asking what changes in the present would be required to arrive at envisioned alternative future states. For example, the Policy Reform and Eco-Communalism scenarios developed by the Global Scenario Group rely on the backcasting method. Practitioners of futures studies classify themselves as futurists (or *foresight practitioners*).

Futurists use a diverse range of forecasting methods including:

- Anticipatory thinking protocols:
- Causal layered analysis (CLA)
- Environmental scanning
- Scenario method
- Delphi method
- Future history
- Monitoring
- Backcasting (eco-history)
- Cross-impact analysis
- Futures workshops
- Failure mode and effects analysis
- Futures wheel
- Technology roadmapping
- Social network analysis
- Systems engineering
- Trend analysis
- Morphological analysis
- Technology forecasting

## 15.4.2  Shaping alternative futures

Futurists use scenarios – alternative possible futures – as an important tool. To some extent, people can determine what they consider probable or desirable using qualitative and quantitative methods. By looking at a variety of possibilities one comes closer to shaping the future, rather than merely predicting it. Shaping alternative futures starts by establishing a number of scenarios. Setting up scenarios takes place as a process with many stages. One of those stages involves the study of trends. A trend persists long-term and long-range; it affects many societal groups, grows slowly and appears to have a profound basis. In contrast, a fad operates in the short term, shows the vagaries of fashion, affects particular societal groups, and spreads quickly but superficially.

Sample predicted futures range from predicted ecological catastrophes, through a utopian future where the poorest human being lives in what present-day observers would regard as wealth and comfort, through the transformation of humanity into a posthuman life-form, to the destruction of all life on Earth in, say, a nanotechnological disaster.

Futurists have a decidedly mixed reputation and a patchy track record at successful prediction. For reasons of convenience, they often extrapolate present technical and societal trends and assume they will develop at the same rate into the future; but technical progress and social upheavals, in reality, take place in fits and starts and in different areas at different rates.

Many 1950s futurists predicted commonplace space tourism by the year 2000, but ignored the possibilities of ubiquitous, cheap computers. On the other hand, many forecasts have portrayed the future with some degree of accuracy. Current futurists often present multiple scenarios that help their audience envision what "may" occur instead of merely "predicting the future". They claim that understanding potential scenarios helps individuals and organizations prepare with flexibility.

Many corporations use futurists as part of their risk management strategy, for horizon scanning and emerging issues analysis, and to identify wild cards – low probability, potentially high-impact risks.[28] Every successful and unsuccessful business engages in futuring to some degree – for example in research and development, innovation and market research, anticipating competitor behavior and so on.[29][30]

## 15.4.3  Weak signals, the future sign and wild cards

In futures research "weak signals" may be understood as advanced, noisy and socially situated indicators of change in trends and systems that constitute raw informational material for enabling anticipatory action. There is some confusion about the definition of weak signal by various researchers and consultants. Sometimes it is referred as future oriented information, sometimes more like emerging issues. The confusion has been partly clarified with the concept 'the future sign', by separating signal, issue and interpretation of the future sign.[31]

"Wild cards" refer to low-probability and high impact events, such as existential risks. This concept may be embedded in standard foresight projects and introduced into anticipatory decision-making activity in order to increase the ability of social groups adapt to surprises arising in turbulent business environments. Such sudden and unique incidents might constitute turning points in the evolution of a certain trend or system. Wild cards may or may not be announced by weak signals, which are incomplete and fragmented data from which relevant foresight information might be inferred. Sometimes, mistakenly, wild cards and weak signals are considered as synonyms, which they are not.[32]

## 15.4.4  Near-term predictions

A long-running tradition in various cultures, and especially in the media, involves various spokespersons making predictions for the upcoming year at the beginning of the year. These predictions sometimes base themselves on current trends in culture (music, movies, fashion, politics); sometimes they make hopeful guesses as to what major events might take place over the course of the next year.

Some of these predictions come true as the year unfolds, though many fail. When predicted events fail to take place, the authors of the predictions often state that misinterpretation of the "signs" and portents may explain the failure of the prediction.

Marketers have increasingly started to embrace futures studies, in an effort to benefit from an increasingly competitive marketplace with fast production cycles, using such techniques as trendspotting as popularized by Faith Popcorn.

### 15.4.5 Trend analysis and forecasting

**Mega-trends**

Trends come in different sizes. A mega-trend extends over many generations, and in cases of climate, mega-trends can cover periods prior to human existence. They describe complex interactions between many factors. The increase in population from the palaeolithic period to the present provides an example.

**Potential trends**

Possible new trends grow from innovations, projects, beliefs or actions that have the potential to grow and eventually go mainstream in the future.

**Branching trends**

Very often, trends relate to one another the same way as a tree-trunk relates to branches and twigs. For example, a well-documented movement toward equality between men and women might represent a branch trend. The trend toward reducing differences in the salaries of men and women in the Western world could form a twig on that branch.

**Life-cycle of a trend**

When a potential trend gets enough confirmation in the various media, surveys or questionnaires to show that it has an increasingly accepted value, behavior or technology, it becomes accepted as a bona fide trend. Trends can also gain confirmation by the existence of other trends perceived as springing from the same branch. Some commentators claim that when 15% to 25% of a given population integrates an innovation, project, belief or action into their daily life then a trend becomes mainstream.

## 15.5 Education

Education in the field of futures studies has taken place for some time. Beginning in the United States of America in the 1960s, it has since developed in many different countries. Futures education can encourage the use of concepts, tools and processes that allow students to think long-term, consequentially, and imaginatively. It generally helps students to:

1. conceptualise more just and sustainable human and planetary futures.

2. develop knowledge and skills in exploring probable and preferred futures.

3. understand the dynamics and influence that human, social and ecological systems have on alternative futures.

4. conscientize responsibility and action on the part of students toward creating better futures.

Thorough documentation of the history of futures education exists, for example in the work of Richard A. Slaughter (2004),[33] David Hicks, Ivana Milojević[34] and Jennifer Gidley[35][36][37] to name a few.

While futures studies remains a relatively new academic tradition, numerous tertiary institutions around the world teach it. These vary from small programs, or universities with just one or two classes, to programs that incorporate futures studies into other degrees, (for example in planning, business, environmental studies, economics, development studies, science and technology studies). Various formal Masters-level programs exist on six continents. Finally, doctoral dissertations around the world have incorporated futures studies. A recent survey documented approximately 50 cases of futures studies at the tertiary level.[38]

The largest Futures Studies program in the world is at Tamkang University, Taiwan. Futures Studies is a required course at the undergraduate level, with between three to five thousand students taking classes on an annual basis. Housed in the Graduate Institute of Futures Studies is an MA Program. Only ten students are accepted annually in the program. Associated with the program is the *Journal of Futures Studies*.[39]

The longest running Future Studies program in North America was established in 1975 at the University of Houston–Clear Lake.[40] It moved to the University of Houston in 2007 and renamed the degree to Foresight. The program was established on the belief that if history is studied and taught in an academic setting, then so should the future.

As of 2003, over 40 tertiary education establishments around the world were delivering one or more courses in futures studies. The World Futures Studies Federation[41] has a comprehensive survey of global futures programs and courses. The Acceleration Studies Foundation maintains an annotated list of primary and secondary graduate futures studies programs.[42]

Organizations such as Teach The Future also aim to promote future studies in the secondary school curriculum in order to develop structured approaches to thinking about the future in public school students. The rationale is that a sophisticated approach to thinking about, anticipating, and planning for the future is a core skill requirement that every student should have, similar to literacy and math skills.

## 15.6 Futurists

Main article: Futurist

Several authors have become recognized as futurists. They research trends, particularly in technology, and write their observations, conclusions, and predictions. In earlier eras, many futurists were at academic institutions. John McHale, author of *The Future of the Future*, published a 'Futures Directory', and directed a think tank called *The Centre For Integrative Studies* at a university. Futurists have started consulting groups or earn money as speakers, with examples including Alvin Toffler, John Naisbitt and Patrick Dixon. Frank Feather is a business speaker that presents himself as a pragmatic futurist. Some futurists have commonalities with science fiction, and some science-fiction writers, such as Arthur C. Clarke, are known as futurists. In the introduction to *The Left Hand of Darkness*, Ursula K. Le Guin distinguished futurists from novelists, writing of the study as the business of prophets, clairvoyants, and futurists. In her words, "a novelist's business is lying".

A survey of 108 futurists[43] found the following shared assumptions:

1. We are in the midst of a historical transformation. Current times are not just part of normal history.

2. Multiple perspectives are at heart of futures studies, including unconventional thinking, internal critique, and cross-cultural comparison.

3. Consideration of alternatives. Futurists do not see themselves as value-free forecasters, but instead aware of multiple possibilities.

4. Participatory futures. Futurists generally see their role as liberating the future in each person, and creating enhanced public ownership of the future. This is true worldwide.

5. Long term policy transformation. While some are more policy-oriented than others, almost all believe that the work of futures studies is to shape public policy, so it consciously and explicitly takes into account the long term.

6. Part of the process of creating alternative futures and of influencing public (corporate, or international) policy is internal transformation. At international meetings, structural and individual factors are considered equally important.

7. Complexity. Futurists believe that a simple one-dimensional or single-discipline orientation is not satisfactory. Trans-disciplinary approaches that take complexity seriously are necessary. Systems thinking, particularly in its evolutionary dimension, is also crucial.

8. Futurists are motivated by change. They are not content merely to describe or forecast. They desire an active role in world transformation.

9. They are hopeful for a better future as a "strange attractor".

10. Most believe they are pragmatists in this world, even as they imagine and work for another. Futurists have a long term perspective.

11. Sustainable futures, understood as making decisions that do not reduce future options, that include policies on nature, gender and other accepted paradigms. This applies to corporate futurists and the NGO. Environmental sustainability is reconciled with the technological, spiritual and post-structural ideals. Sustainability is not a "back to nature" ideal, but rather inclusive of technology and culture.

## 15.7 Applications of foresight and specific fields

### 15.7.1 General applicability and use of foresight products

Several corporations and government agencies utilize foresight products to both better understand potential risks and prepare for potential opportunities. Several government agencies publish material for internal stakeholders as well as make that material available to broader public. Examples of this include the US Congressional Budget Office long term budget projections,[44] the National Intelligence Center,[45] and the United Kingdom Government Office for Science.[46] Much of this material is used by policy makers to inform policy decisions and government agencies to develop long term plan. Several corporations, particularly those with long product development lifecycles, utilize foresight and future studies products and practitioners in the development of their business strategies. The Shell Corporation is one such entity.[47] Foresight professionals and their tools are increasingly being utilized in both the private and public areas to help leaders deal with an increasingly complex and interconnected world.

### 15.7.2 Fashion and design

Fashion is one area of trend forecasting. The industry typically works 18 months ahead of the current selling season. Large retailers look at the obvious impact of everything from the weather forecast to runway fashion for consumer tastes. Consumer behavior and statistics are also important for a long-range forecast.

Artists and conceptual designers, by contrast, may feel that consumer trends are a barrier to creativity. Many of these 'startists' start micro trends but do not follow trends themselves.

Design is another area of trend forecasting. Foresight and futures thinking are rapidly being adopted by the design industry to insure more sustainable, robust and humanistic products. Design, much like future studies is an interdisciplinary field that considers global trends, challenges and opportunities to foster innovation. Designers are thus adopting futures methodologies including scenarios, trend forecasting, and futures research.

Holistic thinking that incorporates strategic, innovative and anticipatory solutions gives designers the tools necessary to navigate complex problems and develop novel future enhancing and visionary solutions.

The Association for Professional Futurists has also held meetings discussing the ways in which Design Thinking and Futures Thinking intersect.

### 15.7.3 Energy and alternative sources

The future of energy is a complex topic. There are likely not enough new sources of oil in the Earth to make up for escalating demands from China, India, Africa, and other rapidly developing economies, and to replace declining fields, if future energy consumption patterns of these economies mimic the historical patterns of the developed world. Yet as global population saturates and we move further into the Information Age, energy intensity (use of energy per GDP) has been saturating (slowing in rate of growth) or declining in many countries.[48] There are also over a trillion and a half barrels of proven oil reserves in the world,[49] in the hands of owners who want to make sure as much of that oil is sold as possible before we move to alternative sources. They have a strong incentive to keep the price of oil low enough (e.g., below $100 a barrel) to discourage too-rapid emergence of viable alternatives. While many alternative sources of energy exist, their rate of development and their levels of governmental and corporate R&D funding remain slow and low.  Some futurists see a gap looming between the effective end of the Age of Oil and the expected emergence of new energy sources.[50] Others see no such gap ahead, and expect we'll leave the Age of Oil to one led by solar, fusion, natural gas, and other sources with lots of oil still remaining in the ground.  Social values changes, climate change, accelerating computerization and machine productivity, and other factors may increasingly drive us to more sustainable energy sources with significantly fewer environmental and public health costs than oil and coal.[51][52]

### 15.7.4   Imperial cycles and world order

Imperial cycles represent an "expanding pulsation" of "mathematically describable" macro-historic trend.[53] The List of Largest Empires contains imperial record progression in terms of territory or percentage of world population under single imperial rule.

Chinese philosopher K'ang Yu-wei and French demographer Georges Vacher de Lapouge in the late 19th century were the first to stress that the trend cannot proceed indefinitely on the definite surface of the globe.  The trend is bound to culminate in a world empire.  K'ang Yu-wei estimated that the matter will be decided in the contest between Washington and Berlin; Vacher de Lapouge foresaw this contest between the United States and Russia and estimated the chance of the United States higher.[54] Both published their futures studies before H. G. Wells introduced the science of future in his *Anticipations* (1901).

Four later anthropologists—Hornell Hart, Raoul Naroll, Louis Morano, and Robert Carneiro—researched the expanding imperial cycles.  They reached the same conclusion that a world empire is not only pre-determined but close at hand and attempted to estimate the time of its appearance.[55]

Historian Max Ostrovsky, specializing on macro-historic trends and their projection into future, analyzed the inner mechanism at work in the process and applied the results to the conditions of the global system.  The work confirmed the inexorable trend towards a world empire.  He found that the development of the world order in history and its projection into future follows a hyperbolic trajectory.  The research was published in 2007 titled: *Y = Arctg X: The Hyperbola of the World Order.*[56]

### 15.7.5   Education

As Foresight has expanded to include a broader range of social concerns all levels and types of education have been addressed, including formal and informal education.  Many countries are beginning to implement Foresight in their Education policy.  A few programs are listed below:

- Finland's FinnSight 2015[57] - Implementation began in 2006 and though at the time was not referred to as "Foresight" they tend to display the characteristics of a foresight program.

- Singapore's Ministry of Education Master plan for Information Technology in Education[58] - This third Masterplan continues what was built on in the 1st and 2nd plans to transform learning environments to equip students to compete in a knowledge economy.

### 15.7.6   Science fiction

Wendell Bell and Ed Cornish acknowledge science fiction as a catalyst to future studies, conjuring up visions of tomorrow.[59] Science fiction's potential to provide an "imaginative social vision" is its contribution to futures studies and public perspective.  Productive sci-fi presents plausible, normative scenarios.[59] Jim Dator attributes the foundational concepts of "images of the future" to Wendell Bell, for clarifying Fred Polak's concept in Images of the Future, as it applies to futures studies.[60][61] Similar to futures studies' scenarios thinking, empirically supported visions of the future are a window into what the future could be.  Pamela Sargent states, "Science fiction reflects attitudes typical of this century." She gives a brief history of impactful sci-fi publications, like The Foundation Trilogy, by Isaac Asimov and Starship Troopers, by Robert A. Heinlein.[62] Alternate perspectives validate sci-fi as part of the fuzzy "images of the future."[61] However, the challenge is the lack of consistent futures research based literature frameworks.[62] Ian Miles reviews The New Encyclopedia of Science Fiction," identifying ways Science Fiction and Futures Studies "cross-fertilize, as well as the ways in which they differ distinctly." Science Fiction cannot be simply considered fictionalized Futures Studies. It may have aims other than "prediction, and be no more concerned with shaping the future than any other genre of literature." [63] It is not to be understood as an explicit pillar of futures studies, due to its inconsistency of integrated futures research. Additionally, Dennis Livingston, a literature and Futures journal critic says, "The depiction of truly alternative societies has not been one of science fiction's strong points, especially" preferred, normative envisages.[64]

## 15.8  Research centers

- Graduate Degree in Foresight, University of Houston[65]

- Institute for Futures Research, University of Stellenbosch

- Copenhagen Institute for Futures Studies

- The Finland Futures Research Centre (FFRC)

- The Foresight Programme, London, Department for Business, Innovation and Skills

- The Futures Academy, Dublin Institute of Technology, Ireland

- Hawaii Research Center for Futures Studies, University of Hawai'i at Mānoa

- Institute for Futures Research, South Africa

- Institute for the Future, Palo Alto, California

- National Intelligence Council, Office of the Director of National Intelligence, Washington DC

- Machine Intelligence Research Institute (MIRI), Berkeley CA (Previously known as the Singularity Institute )

- Tellus Institute, Boston MA

- World Future Society

- World Futures Studies Federation, world

- Future of Humanity Institute

- Italian Institute for the Future, Naples, Italy[66]

## 15.9  Futurists and foresight thought leaders

Main article: List of futurologists

- Daniel Bell

- Peter C. Bishop

- Nick Bostrom

- Arthur C. Clarke[67]

- Jim Dator

- Leonardo da Vinci (Flight)

- Nicolas De Santis
- Peter Diamandis
- Walt Disney
- Mahdi Elmandjra
- Jacque Fresco[68]
- George Friedman
- Hugo de Garis
- Jennifer M. Gidley
- Ben Goertzel
- Arthur Harkins
- Stephen Hawking[69][70]
- Aldous Huxley ("*Brave New World*")
- Sohail Inayatullah
- Mitchell Joachim
- Bill Joy
- Robert Jungk
- Herman Kahn
- Michio Kaku
- Ray Kurzweil
- Max More
- George Orwell ("*Nineteen Eighty-Four*")
- David Passig
- Kim Stanley Robinson
- Michel Saloff Coste
- Anders Sandberg
- Peter Schwartz
- John Smart
- Mark Stevenson ("*An Optimist's Tour of the Future*")
- Alvin Toffler ("*Future Shock*")
- Jules Verne ("*From the Earth to the Moon*")
- Natasha Vita-More
- H. G. Wells (World Brain)
- Eliezer Yudkowsky

## 15.10 Books

### 15.10.1 Periodicals and monographs

- International Journal of Forecasting

- Journal of Futures Studies

- Technological Forecasting and Social Change

- The Futurist World Future Society

## 15.11 Organizations

## 15.12 See also

- Accelerating change

- Biocultural evolution

- List of emerging technologies

- Human overpopulation

- Outline of future studies

- The Human genetic engineering, cyborg technology, and other hypothetical forms of the future human evolution.

## 15.13 References

[1] "Futurology". *Wordnet Search 3.1*. Princeton University. Retrieved 16 March 2013.

[2] Wells, H.G. (1932) 1987. Wanted: Professors of Foresight! *Futures Research Quarterly* V3N1 (Spring): p. 89-91.

[3] "SCIENCE GLOSSARY". *tripod.com*.

[4] Galtung, Johan and Inayatullah, Sohail (1997). *Macrohistory and Macrohistorians*. Westport, Ct: Praeger.

[5] Khaldun, Ibn (1967), *The Muqaddimah*, Trans. Franz Rosenthal, ed. N.J. Dawood. Princeton: Princeton University Press

[6] Bell, Wendell (1997). *Foundations of Futures Studies: Human Science for a New Era*. New Brunswick, New Jersey, USA: Transaction Publishers. ISBN 1-56000-271-9.

[7] "Samuel Madden's *Memoirs of the Twentieth Century*" Paul Alkon. *Science Fiction Studies* Vol. 12, No. 2 (Jul., 1985), pp. 184-201 Published by: SF-TH Inc

[8] "And now for the forecast". The Guardian.

[9] W. Warren Wagar (1983). "H.G. Wells and the Genesis of Future Studies".

[10] *Anticipations*, p 100-101, 107.

[11] "Annual HG Wells Award for Outstanding Contributions to Transhumanism". Web.archive.org. 20 May 2009. Archived from the original on 20 May 2009. Retrieved 10 June 2012.

[12] Turner, Frank Miller (1993). "Public Science in Britain 1880–1919". *Contesting Cultural Authority: Essays in Victorian Intellectual Life*. Cambridge University Press. pp. 219–20. ISBN 0-521-37257-7.

[13] Richard Rhodes (1986). *The Making of the Atomic Bomb*. New York: Simon & Schuster. p. 24. ISBN 0-684-81378-5.

[14] Cowley, Malcolm. "Outline of Wells's History." *The New Republic* Vol. 81 Issue 1041, 14 November 1934 (p. 22–23).

[15] Masini, Eleonora (1993). *Why Futures Studies?*. London, UK: Grey Seal Books.

[16] Slaughter, Richard A. (1995). *The Foresight Principle: Cultural Recovery in the 21st Century*. London, England: Adamantine Press, Ltd.

[17] Sardar, Ziauddin, ed. (1999). *Rescuing All Our Futures*. Praeger Studies on the 21st Century, Westport, Connecticut, USA.

[18] Meadows, Donella H.; D.L. Meadows, J. Randers, and William W. Behrens III (1972). *The Limits to Growth*. New York, New York, USA: Universe Books.

[19] Markley, Oliver (1998)"Visionary Futures: Guided Imagery in Teaching and Learning about the Future", in *American Behavioral Scientist*. Sage Publications, New York.

[20] Jones, Christopher (Winter 1992). "The Manoa School of Futures Studies". *Futures Research Quarterly*: 19–25.

[21] Kuhn, Thomas (1975, c1970). *The Structure of Scientific Revolutions*. University of Chicago Press, Chicago, Illinois, USA.

[22] Masini, Eleonora (1993). Why Futures Studies?. London, UK: Grey Seal Books.

[23] Dator, James (2002), Advancing Futures, Westport: Ct, Praeger, 2002

[24] Sardar, Ziauddin, ed.,(1999) Rescuing all our futures: the futures of futures studies. Westport, Ct: Praeger

[25] Inayatullah, Sohail (2007), Questioning the Future: methods and tools for organizational and societal change. Tamsui: Tamkang University (third edition)

[26] Slaughter, Richard (2005). *The Knowledge Base of Futures studies*.

[27] Bell, Wendell (1997). *The Foundations of Futures Studies*.

[28] A sample presentation on risk management

[29] Rohrbeck, Rene (2010) *Corporate Foresight: Towards a Maturity Model for the Future Orientation of a Firm*, Springer Series: Contributions to Management Science, Heidelberg and New York, ISBN 978-3-7908-2625-8

[30] Rohrbeck, R. H.G. Gemuenden (2010) Corporate Foresight: Its Three Roles in Enhancing the Innovation Capacity of a Firm" *Technological Forecasting and Social Change*, forthcoming

[31] article about the Future sign

[32] differences of weak signals and wild cards

[33] Slaughter, Richard A. (2004). *Futures Beyond Dystopia: Creating Social Foresight*. London: RoutledgeFalmer.

[34] "Articles by Ivana Milojevic; Futures Studies at Metafuture.org". *metafuture.org*.

[35] Futures in Education: Principles, Practices and Potential, (Monograph No 5, The Strategic Foresight Monograph Series, 2004)]

[36] The University in Transformation: Global Perspectives on the Futures of the University (Westport, Ct., Bergin and Garvey, 2000)

[37] Youth Futures: Empirical Research and Transformative Visions (Westport, Ct. Praeger, 2002)

[38] Super User. "HOME". *wfsf.org*.

[39] "Journal of Future Studies". Tamsui, Taipei, Taiwan.: Graduate Institute of Futures Studies, Tamkang University.

[40] Teaching about the Future, by Peter C. Bishop and Andy Hines, 2012

[41] WFSF Directory of Tertiary Futures Education

[42] "Foresight and Futures Studies – Global Academic Programs". Accelerating.org. 2005-11-04. Retrieved 2009-07-20.

[43] Sohail Inayatullah, ed., The Views of Futurists. Vol 4, The Knowledge Base of Futures Studies. Brisbane, Foresight International, 2001.

[44] "Long-Term Budget Projections". *Congressional Budget Office*.

[45] Super User. "National Intelligence Council". *dni.gov*.

[46] "Foresight projects". *www.gov.uk*.

[47] "Shell Scenarios". *shell.com*.

[48] International Energy Statistics, U.S. Energy Information Administration http://www.eia.gov/cfapps/ipdbproject/iedindex3.cfm?tid=92&pid=46&aid=2

[49] CIA World Factbook https://www.cia.gov/library/publications/the-world-factbook/rankorder/2244rank.html

[50] Hawaii Research Center for Futures Studies, University of Hawai'i at Mānoa. Honolulu Advertiser 2008. http://www.futures.hawaii.edu/publications/energy/DoingLessWithLess2008.pdf

[51] The Bottomless Well, Peter Huber and Mark Mills, Basic Books, 2006. http://www.amazon.com/Bottomless-Well-Twilight-Virtue-Energy/dp/046503117X

[52] Lives Per Gallon, Terry Tamminen, Shearwater, 2008 http://www.amazon.com/Lives-Per-Gallon-True-Addiction/dp/1597265063

[53] Hornell Hart, "The Logistic Growth of Political Areas," *Social Forces*, 26, (1948): 396-7; Raoul Naroll, "Imperial Cycles and World Order," *Peace Research Society*, 7, (1967): 100-101.

[54] K'ang Yu-wei, *The One World Philosophy*, (tr. Thompson, Lawrence G., London, 1958), pp 79-80, 85; George Vacher de Lapouge, *L'Aryen: Son Rôle Social*, (Nantes: 1899), chapter " L`Avenir des Aryens."

[55] Hornell, Hart, "The Logistic Growth of Political Areas," *Social Forces*, 26, (1948): 396-408; Raoul, Naroll, "Imperial Cycles and World Order," *Peace Research Society*, 7, (1967): 83-101; Louis A., Marano, "A Macrohistoric Trend Towards World Government", *Behavior Science Notes*, 8, (1973): 35-40; Robert Carneiro, "Political Expansion as an Expression of the Principle of Competitive Exclusion", *Studying War: Anthropological Perspective*, eds. Reyna, Stephen P. & Dawns, Richard Erskine, Gordon and Breach, New Hampshire, 1994; Robert Carneiro, "The Political Unification of the World", *Cross Cultural Survey*, 38/2, (2004), 162-177.

[56] (Lanham: University Press of America).

[57] http://www.aka.fi/Tiedostot/Tiedostot/Julkaisut/Finnsight_2015_EN.pdf

[58] "Ministry of Education, Singapore: Press Releases - MOE Launches Third Masterplan for ICT in Education". *moe.gov.sg*.

[59] Morgan, Matthew J. "On the Fringes: Future Opportunities for Futures Studies." Futures Research Quarterly 19.3 (2003): 5-20. Web. 4 March 2015

[60] Dator, Jim. "Wendell Bell: The Futurist Who Would Put My Grandmother in Prison." Futures 43.6 (2011): 578-82. Web. 4 May 2015

[61] Polak, Fred, and Boulding, Elise. The Image of the Future. (1973). Print.

[62] Women in science fiction. Sargent Pamela. (1975) Futures, 7 (5), pp. 433-441.

[63] Fiction and forecasting. Ian Miles. (1990) Futures, 22 (1) , pp. 83-91

[64] Science Fiction Survey. Dennis Livingston. Futures, Volume 4, Issue 1, March 1972, Pages 97-98

[65] "Graduate Program in Foresight". Retrieved 11 August 2014.

[66] "Italian Institute for the Future". Retrieved 27 October 2014.

[67] "Compressed Data; On a Futurists' Forum, Money Backs Up Predictions", *The New York Times*, April 1, 2002

[68] Fresco, Jacque. "The Venus Project". *The Venus Project: Beyond Politics, Poverty, and War*. Retrieved 16 March 2013.

[69] Nick Paton Walsh. "Alter our DNA or robots will take over, warns Hawking". *the Guardian*.

[70] "BBC NEWS - UK - Move to new planet, says Hawking". *bbc.co.uk*.

## 15.14   External links

- Future at DMOZ

*H. G. Wells first advocated for 'future studies', in a lecture delivered in 1902.*

# Chapter 16

# Futuribles International

**Futuribles International** (formerly **Association Internationale Futuribles**) is a Paris-based international, independent, private non-profit organization network on future studies. It was created in 1960 by Bertrand de Jouvenel while the "Centre d'études prospectives" was created by Gaston Berger in 1957.

The organization main focus is on the impact of new technologies, social policies, lifestyle changes, strategic foresight, and the emergence of an information society.

## 16.1 Publications

- *Revue Futuribles* (journal) (French)

## 16.2 See also

- List of future studies topics
- World Future Society
- World Futures Studies Federation
- Science and technology studies

## 16.3 External links

- Official website (French) and (English)

# Chapter 17

# Ray Hammond

**Ray Hammond** is a British author and futurist.

## 17.1 Selected bibliography

### 17.1.1 Fiction

- *The Cloud* (2006)
- *Emergence* (2002)
- *Extinction* (2005)

### 17.1.2 Non-fiction

- *Forward 100* (1984)
- *Digital Business: Surviving and Thriving In An On-Line World* (1996)
- *The Modern Frankenstein - Fiction Becomes Fact* (1986)
- *The Musician and the Micro* (1983)

## 17.2 External links

- Ray Hammond home page

# Chapter 18

# I = PAT

**I = PAT** is the lettering of a formula put forward to describe the impact of human activity on the environment.

$$I = P \times A \times T$$

In words:

> Human **I**mpact on the environment equals the product of **P**opulation, **A**ffluence, and **T**echnology. This shows how the population, affluence and technology produce an impact

The equation was developed in the 1970s during the course of a debate between Barry Commoner, Paul R. Ehrlich and John Holdren. Commoner argued that environmental impacts in the United States were caused primarily by changes in its production technology following World War II, while Ehrlich and Holdren argued that all three factors were important and emphasized in particular the role of human population growth.[1][2][3]

The equation can aid in understanding some of the factors affecting human impacts on the environment,[4] but it has also been cited as one of the primary factors underlying many of the dire environmental predictions of the 1970s by Paul Ehrlich, George Wald, Denis Hayes, Lester Brown, René Dubos, and Sidney Ripley that did not come to pass.[5] Neal Koblitz classified equations of this type as "mathematical propaganda" and criticized Ehrlich's use of them in the media (e.g. on The Tonight Show) to sway the general public.[6]

The Kaya identity is closely related to the I = PAT equation. The I = PAT equation is more general, describing an abstract "impact". The Kaya identity describes more clearly the impact of human activity on CO2 emissions.

## 18.1   Population

See also: World Population and Human overpopulation

In the **I=PAT** equation, the variable **P** represents the population of an area, such as the world. Since the rise of industrial societies, human population has been increasing exponentially. This has caused Thomas Malthus and many others to postulate that this growth would continue until checked by widespread hunger and famine (see Malthusian growth model).

The United Nations and the US Census Bureau project that world population will increase from 7.0 billion today to about 9.2 billion by 2050.[7][8] These projections take into consideration that population growth has slowed in recent years as women are having fewer children. This phenomenon is believed to be a result of demographic transition in developed nations. As a result, the UN believes that human population might stabilize around 9 billion by 2100.[1] However, since the world population is set to keep rising for the next few decades, this factor of the **I=PAT** equation will likely keep increasing human impact on the environment for the near future.

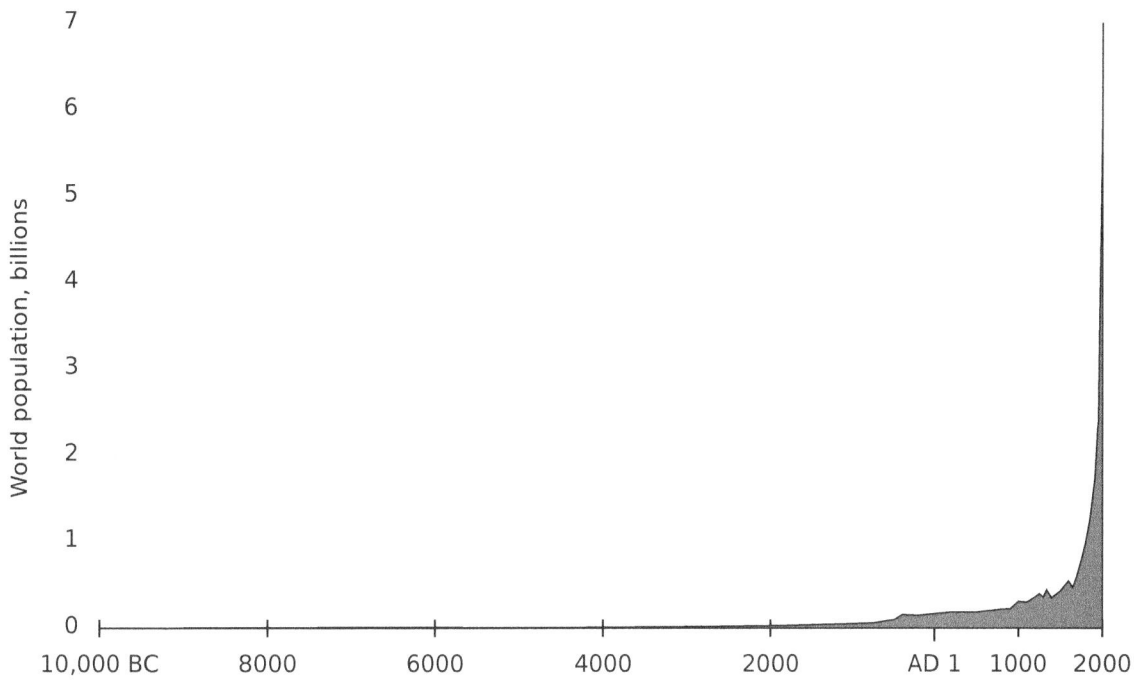

*Population (est.) 10,000 BC – 2000 AD.*

### 18.1.1 Environmental Impacts

Increased population increases humans' environmental impact in many ways, which include but are not limited to:

- Increased land use - Results in habitat loss for other species.

- Increased resource use - Results in changes in land cover

- Increased pollution - Can cause sickness and damages ecosystems.

## 18.2 Affluence

Main article: Consumption (economics)
 The variable **A**, in the **I=PAT** equation stands for affluence. It represents the average consumption of each person in the population. As the consumption of each person increases, the total environmental impact increases as well. A common proxy for measuring consumption is through GDP per capita. While GDP per capita measures production, it is often assumed that consumption increases when production increases. GDP per capita has been rising steadily over the last few centuries and is driving up human impact in the **I=PAT** equation.

### 18.2.1 Environmental Impacts

Increased consumption significantly increases human environmental impact. This is because each product consumed has wide ranging effects on the environment. For example, if the construction of a car had the following environmental impacts among others:

- 605,664 gallons of water for parts and tires[9]

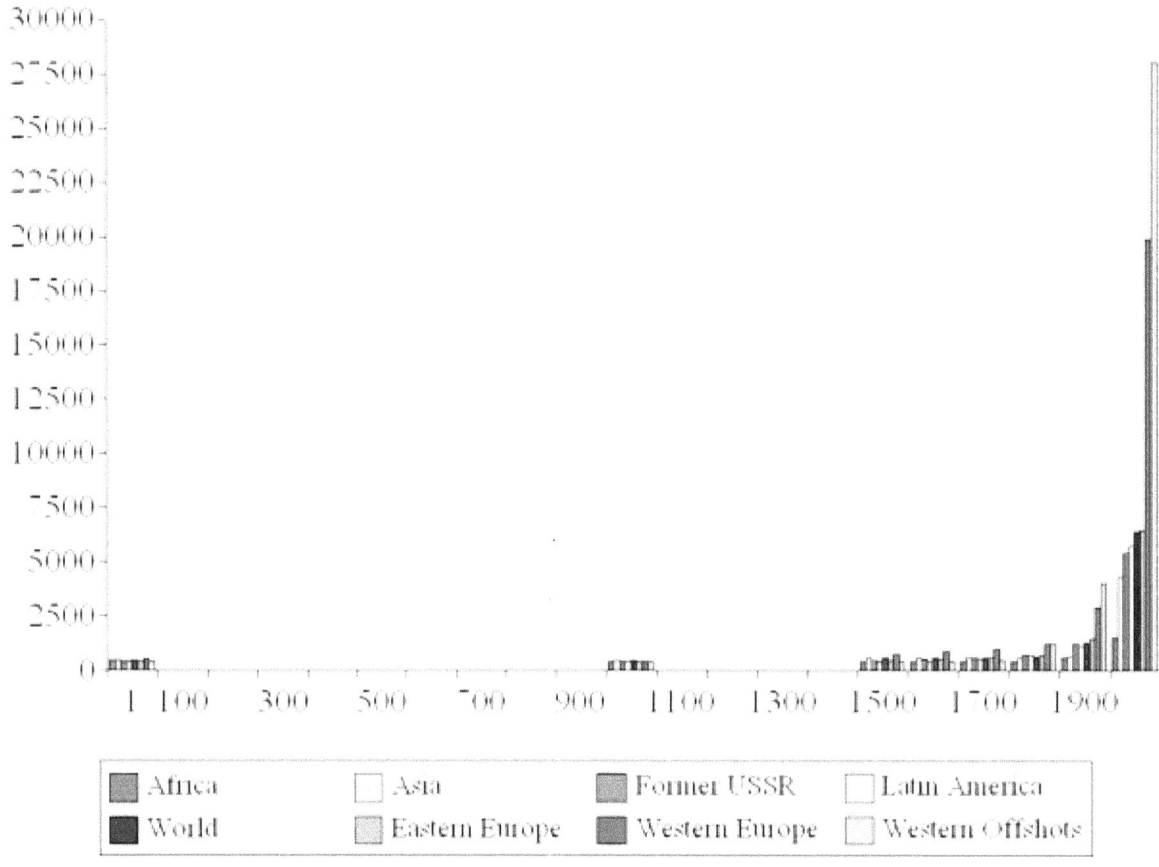

*World GDP per capita (in 1990 Geary-Khamis dollars)*

- 682 lbs. of pollution at a mine for the lead battery.[9]

- 2178 lbs. of discharge into water supply for the 22 lbs. of copper contained in the car.[9]

then the more cars per capita, the greater the impact. Since the ecological impacts of each product are far reaching, increases in consumption quickly result in large impacts on the environment.

## 18.3  Technology

See also: Environmental technology and Appropriate technology

The **T** variable in the **I=PAT** equation represents how resource intensive the production of affluence is; how much environmental impact is involved in creating, transporting and disposing of the goods, services and amenities used. Improvements in efficiency can reduce resource intensiveness, reducing the T multiplier. Since technology can affect environmental impact in many different ways, the unit for **T** is often tailored for the situation **I=PAT** is being applied to. For example, for a situation where the human impact on climate change is being measured, an appropriate unit for **T** might be greenhouse gas emissions per unit of GDP.

### 18.3.1 Environmental Impacts

Increases in efficiency can reduce overall environmental impact. However, since **P** has increased exponentially, and **A** has also increased drastically, the overall environmental impact, **I**, has still increased.

## 18.4 Reception

The I=PAT equation has been criticized for being too simplistic by assuming that P, A, and T are independent of each other. In reality, at least 7 interdependencies between P, A, and T could exist, indicating that it is more correct to rewrite the equation as I = f(P,A,T).[10] For example, a doubling of technological efficiency, or equivalently a reduction of the T-factor by 50%, does not necessarily reduce the environmental impact (I) by 50% if efficiency induced price reductions stimulate additional consumption of the resource that was supposed to be conserved, a phenomenon called the rebound effect (conservation) or Jevons Paradox. As was shown by Alcott,[10]:Fig. 5 despite significant improvements in the carbon intensity of GDP (i.e., the efficiency in carbon use) since 1980, world fossil energy consumption has increased in line with economic and population growth. Similarly, an extensive historical analysis of technological efficiency improvements has conclusively shown that energy and materials use efficiency improvements were almost always outpaced by economic growth, resulting in a net increase in resource use and associated pollution.[11][12]

As a result of the interdependencies between P, A, and T and potential rebound effects, policies aimed at decreasing environmental impacts through reductions in P, A, and T may not only be very difficult to implement (i.e., population control and material sufficiency and degrowth movements have been very controversial) but also are likely to be rather ineffective compared to rationing (i.e., quotas) or Pigouvian taxation of resource use or pollution.[10]

## 18.5 See also

- Affluence

- Carbon footprint

- Ecological footprint

- Ecological indicator

- Embodied energy

- Life cycle assessment

- Sustainability measurement

- Sustainability metrics and indices

- Technology

- Water footprint

## 18.6 References

[1] O'Neill, B.C.; MacKellar, F.L.; Lutz, W. (2004). "Population, greenhouse gas emissions, and climate change". In Lutz, W.; Sanderson, W.C.; Scherbov, S. *The End of World Population Growth in the 21st Century: New Challenges for Human Capital Formation & Sustainable Development*. London: Earthscan Press. pp. 283–314.

[2] Ehrlich, Paul R.; Holdren, John P. (1971). "Impact of Population Growth". *Science* (American Association for the Advancement of Science) **171** (3977): 1212–1217. doi:10.1126/science.171.3977.1212. JSTOR 1731166.

[3]  Barry Commoner (May 1972). "A Bulletin Dialogue: on "The Closing Circle" - Response". *Bulletin of the Atomic Scientists*:
     17–56.

[4]  Chertow, M. R. (2000). "The IPAT Equation and Its Variants". *Journal of Industrial Ecology* **4** (4): 13–29. doi:10.1162/108819800525419277.

[5]  R Bailey (2000) *Earth day then and now*, Reason **32**(1), 18-28

[6]  N Koblitz (1981) "Mathematics as Propaganda", in *Mathematics Tomorrow*, ed. Lynn Steen, pp 111-120.

[7]  US Census Bureau international popopulation statistics and projections 1950 to 2050

[8]  United Nations population projections

[9]  Andriantiatsaholiniaina, L. A.; Kouikoglou, V. S.; Phillis, Y. A. (2004). "Evaluating strategies for sustainable development:
     Fuzzy logic reasoning and sensitivity analysis". *Ecological Economics* **48** (2): 149. doi:10.1016/j.ecolecon.2003.08.009.

[10] Alcott, B. (2010). "Impact caps: Why population, affluence and technology strategies should be abandoned". *Journal of Cleaner
     Production* **18** (6): 552–560. doi:10.1016/j.jclepro.2009.08.001.

[11] Huesemann, Michael H., and Joyce A. Huesemann (2011). *Technofix: Why Technology Won't Save Us or the Environment*,
     Chapter 5, "In Search of Solutions II: Efficiency Improvements", New Society Publishers, Gabriola Island, British Columbia,
     Canada, ISBN 0865717044, 464 pp.

[12] Cleveland, C. J.; Ruth, M. (1998). "Indicators of Dematerialization and the Materials Intensity of Use". *Journal of Industrial
     Ecology* **2** (3): 15. doi:10.1162/jiec.1998.2.3.15.≥

# Chapter 19

# Immersion (virtual reality)

**Immersion** into virtual reality is a perception of being physically present in a non-physical world. The perception is created by surrounding the user of the VR system in images, sound or other stimuli that provide an engrossing total environment.

The name is a metaphoric use of the experience of submersion applied to representation, fiction or simulation. Immersion can also be defined as the state of consciousness where a "visitor" (Maurice Benayoun) or "immersant" (Char Davies)'s awareness of physical self is transformed by being surrounded in an artificial environment; used for describing partial or complete suspension of disbelief, enabling action or reaction to stimulations encountered in a virtual or artistic environment. The degree to which the virtual or artistic environment faithfully reproduces reality determines the degree of suspension of disbelief. The greater the suspension of disbelief, the greater the degree of presence achieved.

## 19.1 Types of immersion

According to Ernest W. Adams, author and consultant on game design,[1] immersion can be separated into three main categories:

**Tactical immersion** Tactical immersion is experienced when performing tactile operations that involve skill. Players feel "in the zone" while perfecting actions that result in success.

**Strategic immersion** Strategic immersion is more cerebral, and is associated with mental challenge. Chess players experience strategic immersion when choosing a correct solution among a broad array of possibilities.

**Narrative immersion** Narrative immersion occurs when players become invested in a story, and is similar to what is experienced while reading a book or watching a movie.

Staffan Björk and Jussi Holopainen, in *Patterns In Game Design*,[2] divide immersion into similar categories, but call them **sensory-motoric immersion**, **cognitive immersion** and **emotional immersion**, respectively. In addition to these, they add a new category:

**Spatial immersion** Spatial immersion occurs when a player feels the simulated world is perceptually convincing. The player feels that he or she is really "there" and that a simulated world looks and feels "real".

## 19.2 Presence

Virtual reality glasses can produce a visceral feeling of being in a simulated world, a form of spatial immersion called Presence. According to Oculus VR, the technology requirements to achieve this visceral reaction are low-latency and

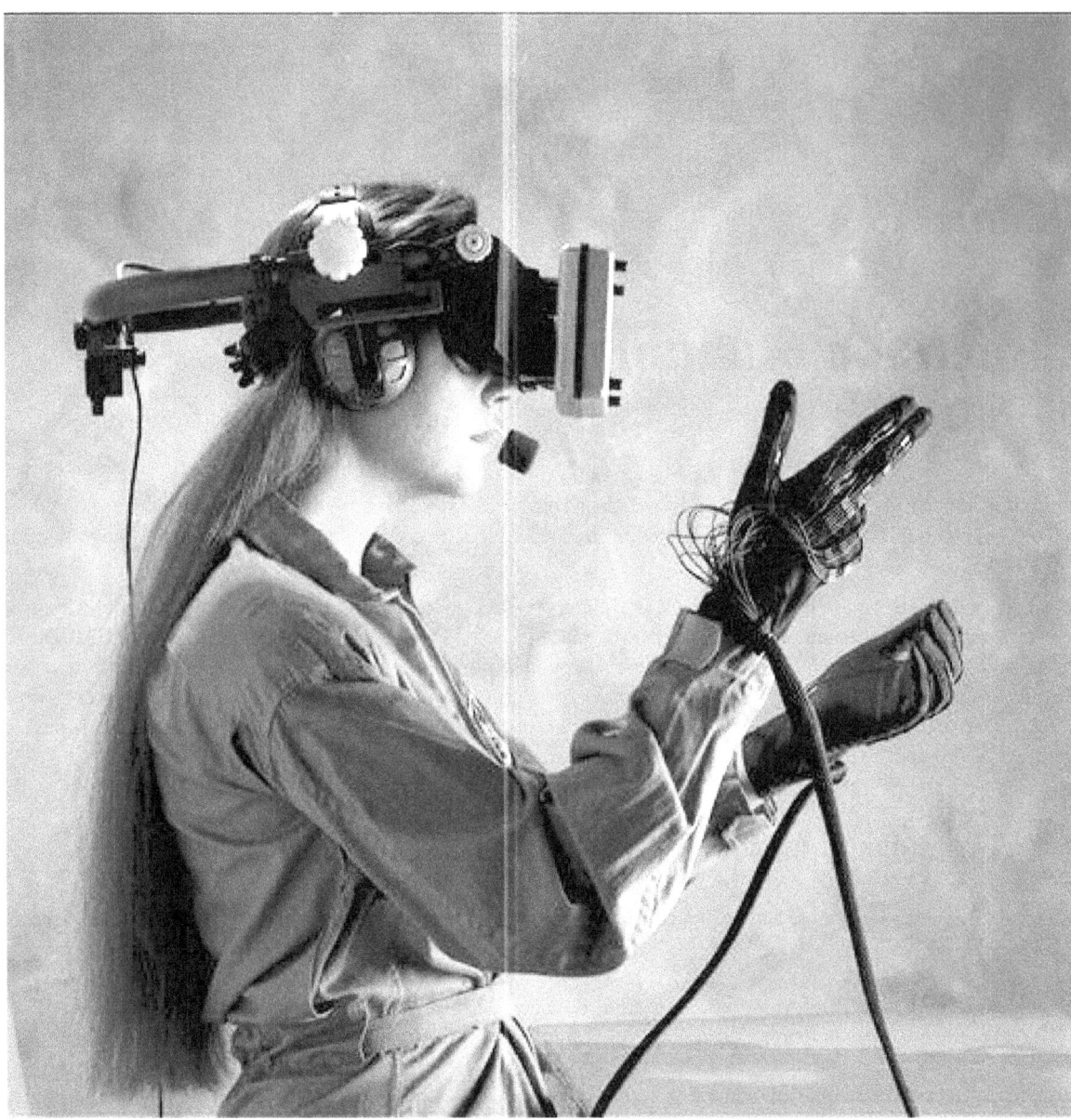

*Classic Virtual reality HMD*

precise tracking of movements.[3][4][5]

Michael Abrash gave a talk on VR at Steam Dev Days in 2014.[6] According to the VR research team at Valve, all of the following are needed to establish presence.

- A wide field of view (80 degrees or better)

- Adequate resolution (1080p or better)

- Low pixel persistence (3 ms or less)

- A high enough refresh rate (>60 Hz, 95 Hz is enough but less may be adequate)

- Global display where all pixels are illuminated simultaneously (rolling display may work with eye tracking.)

- Optics (at most two lenses per eye with trade-offs, ideal optics not practical using current technology)

- Optical calibration

- Rock-solid tracking – translation with millimeter accuracy or better, orientation with quarter degree accuracy or better, and volume of 1.5 meter or more on a side

- Low latency (20 ms motion to last photon, 25 ms may be good enough)

# 19.3 Immersive virtual reality

*The Cave Automatic Virtual Environment*

**Immersive virtual reality** is a hypothetical future technology that exists today as virtual reality art projects, for the most part.[7] It consists of immersion in an artificial environment where the user feels just as immersed as they usually feel in consensus reality.

## 19.3.1 Direct interaction of the nervous system

The most considered method would be to induce the sensations that made up the virtual reality in the nervous system directly. In functionalism/conventional biology we interact with consensus reality through the nervous system. Thus we receive all input from all the senses as nerve impulses. It gives your neurons a feeling of heightened sensation. It would involve the user receiving inputs as artificially stimulated nerve impulses, the system would receive the CNS outputs (natural nerve impulses) and process them allowing the user to interact with the virtual reality. Natural impulses between the body and central nervous system would need to be prevented. This could be done by blocking out natural impulses using nanorobots which attach themselves to the brain wiring, whilst receiving the digital impulses of which describe the virtual world, which could then be sent into the wiring of the brain. A feedback system between the user and the computer which stores the information would also be needed. Considering how much information would be required for such a system, it is likely that it would be based on hypothetical forms of computer technology.

### 19.3.2   Requirements

#### Understanding of the nervous system

A comprehensive understanding of which nerve impulses correspond to which sensations, and which motor impulses correspond to which muscle contractions will be required. This will allow the correct sensations in the user, and actions in the virtual reality to occur. The Blue Brain Project is the current, most promising research with the idea of understanding how the brain works by building very large scale computer models.

#### Ability to manipulate CNS

The nervous system would obviously need to be manipulated. Whilst non-invasive devices using radiation have been postulated, invasive cybernetic implants are likely to become available sooner and be more accurate. Manipulation could occur at any stage of the nervous system – the spinal cord is likely to be simplest; as all nerves pass through here, this could be the only site of manipulation. Molecular Nanotechnology is likely to provide the degree of precision required and could allow the implant to be built inside the body rather than be inserted by an operation.

#### Computer hardware/software to process inputs/outputs

A very powerful computer would be necessary for processing virtual reality complex enough to be nearly indistinguishable from consensus reality and interacting with central nervous system fast enough.

## 19.4   Immersive digital environments

An **immersive digital environment** is an artificial, interactive, computer-created scene or "world" within which a user can immerse themselves.[8]

Immersive digital environments could be thought of as synonymous with Virtual reality, but without the implication that actual "reality" is being simulated. An immersive digital environment could be a model of reality, but it could also be a complete fantasy user interface or abstraction, as long as the user of the environment is immersed within it. The definition of immersion is wide and variable, but here it is assumed to mean simply that the user feels like they are part of the simulated "universe". The success with which an immersive digital environment can actually immerse the user is dependent on many factors such as believable 3D computer graphics, surround sound, interactive user-input and other factors such as simplicity, functionality and potential for enjoyment. New technologies are currently under development which claim to bring realistic environmental effects to the players' environment – effects like wind, seat vibration and ambient lighting.

### 19.4.1   Perception

To create a sense of full immersion, the 5 senses (sight, sound, touch, smell, taste) must perceive the digital environment to be physically real. Immersive technology can perceptually fool the senses through:

- Panoramic 3D displays (visual)

- Surround sound acoustics (auditory)

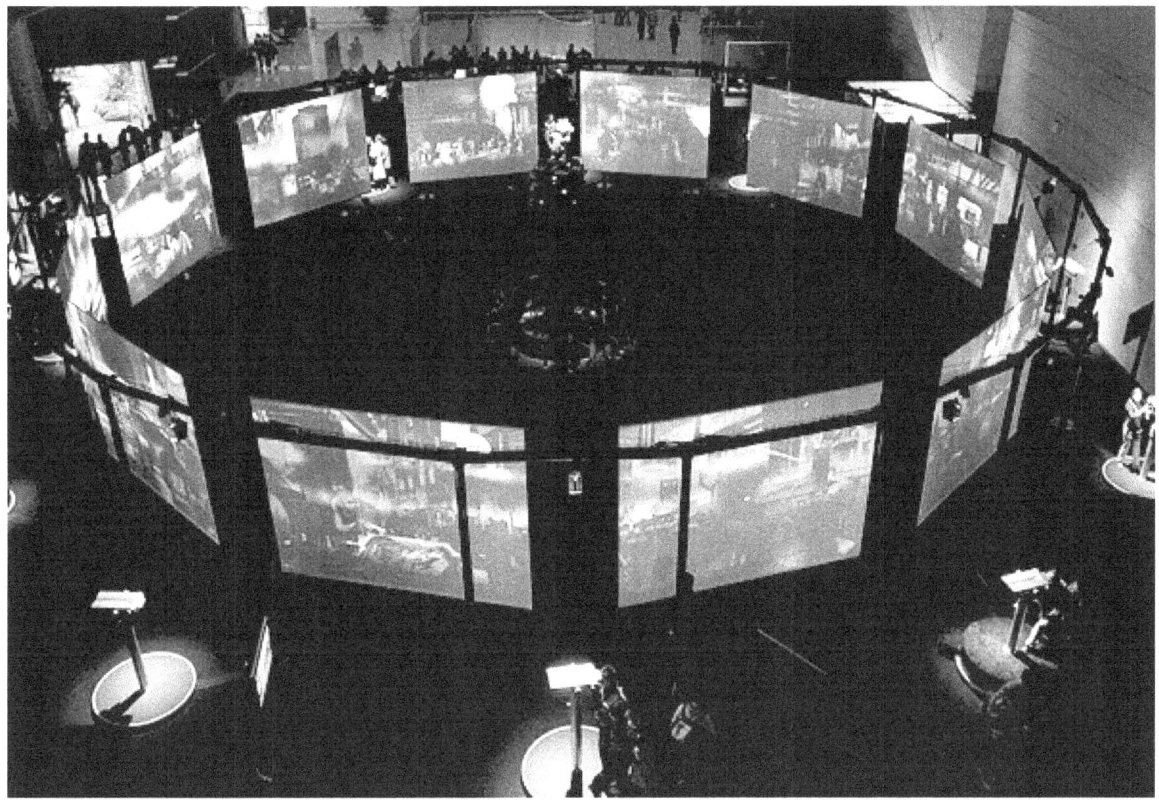

*Cosmopolis (2005), Maurice Benayoun's Giant Virtual Reality Interactive Installation*

- Haptics and force feedback (tactile)
- Smell replication (olfactory)
- Taste replication (gustation)

### 19.4.2  Interaction

Once the senses reach a sufficient belief that the digital environment is real (it is interaction and involvement which can never be real), the user must then be able to interact with the environment in a natural, intuitive manner. Various immersive technologies such as gestural controls, motion tracking, and computer vision respond to the user's actions and movements. Brain control interfaces (BCI) respond to the user's brainwave activity.

### 19.4.3  Examples and applications

Training and rehearsal simulations run the gamut from part task procedural training (often buttonology, for example: which button do you push to deploy a refueling boom) through situational simulation (such as crisis response or convoy driver training) to full motion simulations which train pilots or soldiers and law enforcement in scenarios that are too dangerous to train in actual equipment using live ordinance.

Computer games from simple arcade to Massively multiplayer online game and training programs such as flight and driving simulators. Entertainment environments such as motion simulators that immerse the riders/players in a virtual digital environment enhanced by motion, visual and aural cues. Reality simulators, such as one of the Virunga Mountains in Rwanda that takes you on a trip through the jungle to meet a tribe of Mountain Gorillas.[9] Or training versions such as one which simulates taking a ride through human arteries and the heart to witness the buildup of plaque and thus learn about cholesterol and health.[10]

In parallel with scientist, artists like Knowbotic Research, Donna Cox, Rebecca Allen, Robbie Cooper, Maurice Benayoun, Char Davies, and Jeffrey Shaw use the potential of immersive virtual reality to create physiologic or symbolic experiences and situations.

Other examples of immersion technology include physical environment / immersive space with surrounding digital projections and sound such as the CAVE, and the use of head-mounted displays for viewing movies, with head-tracking and computer control of the image presented, so that the viewer appears to be inside the scene.. The next generation is VIRTSIM, which achieves total immersion through motion capture and wireless head mounted displays for teams of up to thirteen immersants enabling natural movement through space and interaction in both the virtual and physical space simultaneously.

**The use of immersive virtual reality in the medical care**

New fields of studies linked to the immersive virtual reality emerges every day. Researchers see a great potential in virtual reality tests serving as complementary interview methods in psychiatric care.[11] Immersive virtual reality have in studies also been used as an educational tool in which the visualization of psychotic states have been used to get increased understanding of patients with similar symptoms.[12] New treatment methods are available for Schizophrenia [13] and other newly developed research areas where immersive virtual reality is expected to achieve melioration is in education of surgical procedures,[14] rehabilitation program from injuries and surgeries [15] and reduction of phantom limb pain.[16]

## 19.5   Detrimental Issues

Simulation sickness, or simulator sickness, is a condition where a person exhibits symptoms similar to motion sickness caused by playing computer/simulation/video games (Oculus Rift is working to solve simulator sickness).[17]

Motion sickness due to virtual reality is very similar to simulation sickness and motion sickness due to films. In virtual reality, however, the effect is made more acute as all external reference points are blocked from vision, the simulated images are three-dimensional and in some cases stereo sound that may also give a sense of motion. Studies have shown that exposure to rotational motions in a virtual environment can cause significant increases in nausea and other symptoms of motion sickness (So, R.H.Y. and Lo, W.T. (1999) "Cybersickness: An Experimental Study to Isolate the Effects of Rotational Scene Oscillations." Proceedings of IEEE Virtual Reality '99 Conference, March 13–17, 1999, Houston, Texas. Published by IEEE Computer Society, pp. 237–241)

Other behavioural changes such as stress, addiction, isolation and mood changes are also discussed to be side-effects caused by immersive virtual reality. [18]

## 19.6   See also

- Alternate reality game

- Cave automatic virtual environment

- Environmental sculpture

- Escapism

- Immersive design

- Immersive technology

- Interactive art

- Motion sickness

- Narrative transportation

- Neo-conceptual art

- Simulated reality

- Simulation sickness

- Sound art

- Sound installation

- Video installation

- Virtual art

## 19.7 Footnotes

[1] Adams, Ernest (July 9, 2004). "Postmodernism and the Three Types of Immersion". Gamasutra. Retrieved 2007-12-26.

[2] Björk, Staffan; Jussi Holopainen (2004). *Patterns In Game Design*. Charles River Media. p. 206. ISBN 1-58450-354-8.

[3] Oculus Rift Dev Kit 2 now on sale for $350

[4] Oculus Rift DK2 hands-on and first-impressions

[5] Announcing the Oculus Rift Development Kit 2 (DK2)

[6] Abrash M. (2014). *What VR could, should, and almost certainly will be within two years*

[7] Joseph Nechvatal, *Immersive Ideals / Critical Distances*. LAP Lambert Academic Publishing. 2009, pp. 367-368

[8] Joseph Nechvatal, *Immersive Ideals / Critical Distances*. LAP Lambert Academic Publishing. 2009, pp. 48-60

[9] pulseworks.com

[10] usagainstathero.com

[11] Freeman, D.; Antley, A.; Ehlers, A.; Dunn, G.; Thompson, C.; Vorontsova, N.; Garety, P.; Kuipers, E.; Glucksman, E.; Slater, M. (2014). "The use of immersive virtual reality (VR) to predict the occurrence 6 months later of paranoid thinking and posttraumatic stress symptoms assessed by self-report and interviewer methods: A study of individuals who have been physically assaulted". *Psychological Assessment* **26** (3): 841. doi:10.1037/a0036240.

[12] http://www.life-slc.org/docs/Bailenson_etal-immersiveVR.pdf

[13] Freeman, D. (2007). "Studying and Treating Schizophrenia Using Virtual Reality: A New Paradigm". *Schizophrenia Bulletin* **34** (4): 605. doi:10.1093/schbul/sbn020.

[14] *Virtual Reality in Neuro-Psycho-Physiology*, p. 36, at Google Books

[15] De Los Reyes-Guzman, A.; Dimbwadyo-Terrer, I.; Trincado-Alonso, F.; Aznar, M. A.; Alcubilla, C.; Pérez-Nombela, S.; Del Ama-Espinosa, A.; Polonio-López, B. A.; Gil-Agudo, Á. (2014). "A Data-Globe and Immersive Virtual Reality Environment for Upper Limb Rehabilitation after Spinal Cord Injury". *XIII Mediterranean Conference on Medical and Biological Engineering and Computing 2013*. IFMBE Proceedings **41**. p. 1759. doi:10.1007/978-3-319-00846-2_434. ISBN 978-3-319-00845-5.

[16] Llobera, J.; González-Franco, M.; Perez-Marcos, D.; Valls-Solé, J.; Slater, M.; Sanchez-Vives, M. V. (2012). "Virtual reality for assessment of patients suffering chronic pain: A case study". *Experimental Brain Research* **225**: 105. doi:10.1007/s00221-012-3352-9.

[17] "Oculus Rift is working to solve simulator sickness". Polygon. Retrieved 2015-05-05.

[18] http://www.agocg.ac.uk/reports/virtual/37/37.pdf

# 19.8    References

- media.ford.com

- on.aol.com

- Reyes, Stephanie. "Ford brings virtual reality presentation to UCF.", September 2012.

- Christiane Paul, *Digital Art*, Thames & Hudson Ltd.

- Oliver Grau, "Virtual Art: From Illusion to Immersion" MIT-Press, Cambridge 2003

- Timothy Murray, Derrick de Kerckhove, Oliver Grau, Kristine Stiles, Jean-Baptiste Barrière, Dominique Moulon, *Maurice Benayoun Open Art*, Nouvelles éditions Scala, 2011, French version, ISBN 978-2-35988-046-5

- Allen Varney, (August 8, 2006). "Immersion Unexplained" in "The Escapist"

- Frank Popper, "From Technological to Virtual Art", MIT Press. ISBN 0-262-16230-X.

- Oliver Grau (Ed.), *Media Art Histories*, MIT-Press, Cambridge 2007

- Joseph Nechvatal, "Immersive Excess in the Apse of Lascaux", Technonoetic Arts 3, no3. 2005

- Adams, Ernest (July 9, 2004). "Postmodernism and the Three Types of Immersion". Gamasutra. Retrieved 2007-12-26.

- Björk, Staffan; Jussi Holopainen (2004). *Patterns In Game Design*. Charles River Media. p. 423. ISBN 1-58450-354-8.

- Edward A. Shanken, *Art and Electronic Media*. London: Phaidon, 2009. ISBN 978-0-7148-4782-5

- Joseph Nechvatal *Towards an Immersive Intelligence: Essays on the Work of Art in the Age of Computer Technology and Virtual Reality (1993–2006)*. Edgewise Press. New York, N.Y. 2009

- Joseph Nechvatal, *Immersive Ideals / Critical Distances*. LAP Lambert Academic Publishing. 2009

# 19.9    External links

- Annual Summit on Immersive Technology

-  pdf download of Joseph Nechvatal's text book: *Immersive Ideals / Critical Distances*. LAP Lambert Academic Publishing. 2009

- Audio and Game Immersion PhD thesis about game audio (the IEZA Framework) and immersion.

- "Improvising Synesthesia: Comprovisation of Generative Graphics and Music" by Joshua B. Mailman, in *Leonardo Electronic Almanac* v.19 no.3, *Live Visuals*, 2013, pp. 352–84, about two immersive systems for improvising music and graphics through dance-like motion detected by an infrared video camera and other sensors.

# Chapter 20

# Industrial Big Data

Industrial Big Data refers to huge data sets generated by industrial machines, processes or operations. Every unit in an industrial system generates vast amount of data every moment. Billions of data samples are being generated by every single machine per day in a manufacturing line.[1] As another example a Boeing 787 generates over half a terabyte of data per flight.[2] Clearly the volume of data generated by group of units in an industrial system is far beyond the capability of traditional methods therefore handling, managing and processing it would be a challenge.

## 20.1  Sample Repositories

In the course of last several years, researchers and companies have actively participated in collecting, organizing and analyzing huge industrial data sets. Some of these data sets are currently available for public usage for research purposes. NASA data repository[3] is one of the most famous data repositories for Industrial Big Data. Various data sets provided by this repository may be used for predictive analysis, fault detection, prognostics and etc.

## 20.2  References

[1] "The Rise of Industrial Big Data" (PDF).

[2] "Computer World".

[3] "NASA Data Repository". NASA.

[4] "Best Lab at UC Berkeley".

[5] "IMS Center".

[6] "FEMO Institue".

# Chapter 21

# Industrial Internet

The **industrial internet** is a term coined by Frost & Sullivan[1] and refers to the integration of complex physical machinery with networked sensors and software. The industrial Internet draws together fields such as machine learning, big data, the Internet of things, machine-to-machine communication and Cyber-physical system to ingest data from machines, analyze it (often in real-time), and use it to adjust operations.

As of 27 March 2014, the Industrial Internet Consortium (IIC) was founded by AT&T, Cisco, General Electric, IBM, and Intel to bring together industry players—from multinational corporations to academia and governments—to accelerate the development, adoption and wide-spread use of Industrial Internet technologies.[2]

## 21.1 Design Guidelines

Since the meaning of Industrial Internet is similar to that of Cyber-physical Systems (CPS), the design of an Industrial Internet platform can also follow the "5C Architecture".[3] "5C" refers to the five levels of designing a CPS, which is much more clear and concrete than the commonly referenced two functional components: advanced connectivity and intelligent data analytics. These five levels are (1) Smart connection; (2) Data-to-information conversion; (3) Cyber; (4) Cognition; and (5) Configuration.

### 21.1.1 Smart Connection

A necessity of building connection between the cyber space and the physical space is the acquisition of data from industrial equipment. Data might be collected from different sources: add-on sensors, controllers, human inspection, maintenance log, alarm / event systems, etc. One of the challenges at this level is the diversity of equipment types and communication protocols. Machine-to-Machine techniques such as MTConnect[4] are of vital importance to serve as one of the solutions.

### 21.1.2 Data-to-Information Conversion

The era of Big Data does not promise self-evident insightful information. Instead, it usually means that significant effort has to be taken to "mine" knowledge among a large number of less useful information. Through the development of machine learning, statistical analysis, and data mining techniques, this level of the platform will perform intelligent analysis on the data and bring self-awareness to assets.

### 21.1.3 Cyber

Cyber level is what fundamentally differentiates CPS or Industrial Internet from conventional data-driven modeling frameworks. The cyber level serves as a central information hub, where information from networked machines or assets is

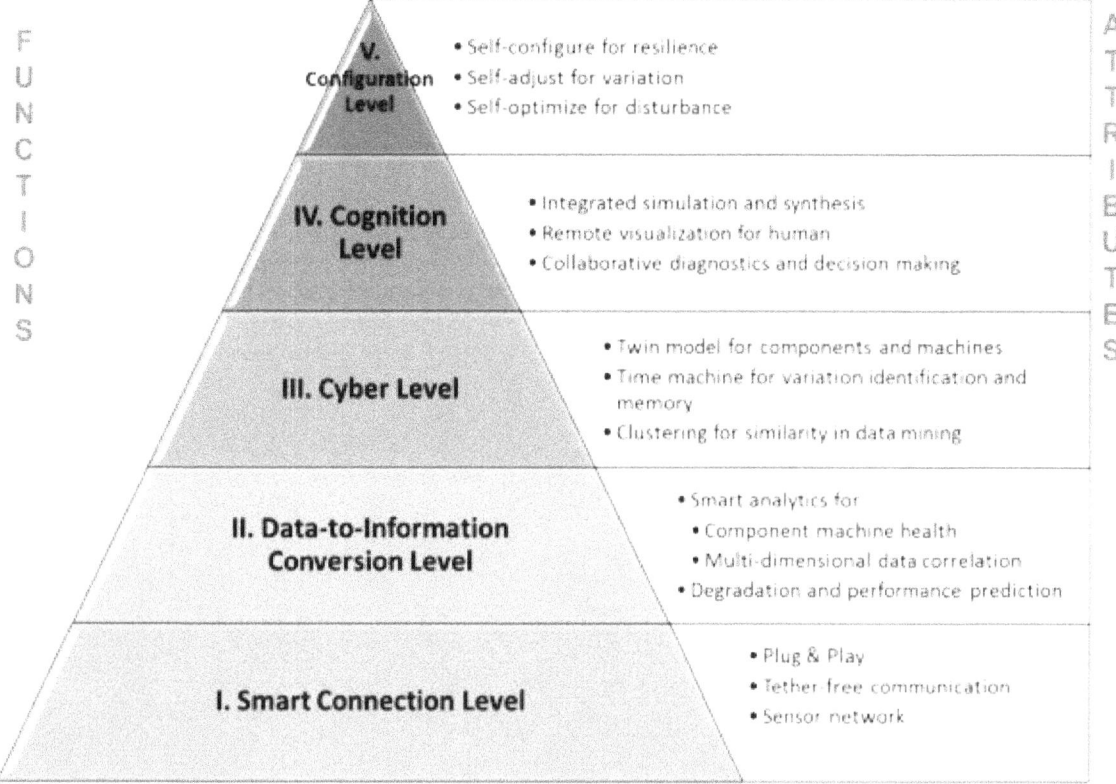

pumped into it, and customized analytics will be performed to extract knowledge of machine conditions over time. These analytics will equip machines with the ability of self-comparison – the foundation of learning from the past. On the other hand, similarities between machine performances and other units can be measured to predict the future behavior – the ability of peer comparison, which brings more accurate prediction.

### 21.1.4  Cognition

The task of Cognition level is to generate a thorough understanding of the monitored machines or assets, so that the acquired insights can be visualized by users to better support decision-making.

### 21.1.5  Configuration

Eventually, Configuration level will realize the feedback from Cyber space to the Physical space. Resilience control system (RCS) methodology will be applied to make corrective and preventive decisions to the monitored system.

## 21.2  Software platform

**Predix** is General Electrics's software platform for the Industrial Internet.[5]

## 21.3   Examples

The Google driverless car takes in environmental data from roof-mounted LIDAR, uses machine-vision techniques to identify road geometry and obstacles, and controls the car's throttle, brakes and steering mechanism in real-time.[6]

The Union Pacific Railroad mounts infrared thermometers, microphones and ultrasound scanners alongside its tracks. These sensors scan every train as it passes and send readings to the railroad's data centers, where pattern-matching software identifies equipment at risk of failure.[7][8] Falling prices for computing power and networked sensors mean that similar techniques can be applied to small, common devices like machine tools.[9] In that case scenario, following a 5C architecture defined for Cyber-Physical Systems will help to standardize the use of Industrial Internet in manufacturing and related disciplines[3][10]

## 21.4   See also

- Cloud-based design and manufacturing

- Big data

- SCADA

- Industrial Ethernet

- Internet of Things

- Machine to machine

- Industrial control system

- Industry 4.0

- Intelligent Maintenance Systems

- Cyber-physical system

## 21.5   References

[1] "The Industrial Internet". 2000-06-08. Retrieved 2005-06-08.

[2] Hardy, Quentin. "Consortium Wants Standards for Internet of Things". *New York Times*. 27 March 2014.

[3] Lee, Jay; Bagheri, Behrad; Kao, Hung-An (January 2015). "A Cyber-Physical Systems architecture for Industry 4.0-based manufacturing systems". *Manufacturing Letters* **3**: 18–23. doi:10.1016/j.mfglet.2014.12.001.

[4] Vijayaraghavan, Athulan; Sobel, Will; Fox, Armando; Dornfeld, David; Warndorf, Paul (2008-06-23). "Improving machine tool interoperability using standardized interface protocols: MT connect". *Proceedings of 2008 ISFA*.

[5] GE. "Predix Powers Industrial-strength Apps". gesoftware.com. Retrieved 2015-06-29.

[6] Steve Lohr. "The Internet Gets Physical". *The New York Times*. Retrieved 2013-08-18.

[7] Chris Murphy (2012-08-08). "Union Pacific Delivers Internet Of Things Reality Check - Global Cio". Informationweek.com. Retrieved 2013-08-18.

[8] Chris Murphy (2012-12-07). "Silicon Valley Needs To Get Out More - Global Cio - Executive". Informationweek.com. Retrieved 2013-08-18.

[9] Jon Bruner (2012-10-29). "Listening for tired machinery - O'Reilly Radar". Radar.oreilly.com. Retrieved 2013-08-18.

[10] "IMSCenter". *IMSCenter*.

## 21.6   External links

- Mark Fell. "Roadmap for the Internet of Things - Its Impact, Architecture and Future Governance" Carré & Strauss, 2014.

- Lohr, Steve. "The Internet Gets Physical" New York Times, December 17, 2011.

- Bruner, Jon. "Defining the industrial Internet" O'Reilly Radar, January 11, 2013.

- Murphy, Chris. "Silicon Valley Needs To Get Out More" InformationWeek, December 7, 2012.

- Loukides, Mike. "To eat or be eaten?" O'Reilly Radar, November 30, 2012.

- N.P., Ullekh. "How GE's over $100 billion investment in 'industrial internet' will add $15 trillion to world GDP" Economic Times, December 16, 2012.

- Smarr, Larry. "An Evolution Toward a Programmable Universe" New York Times, December 5, 2011.

- Evans, Peter C. and Marco Annunziata. "Industrial Internet: Pushing the Boundaries of Minds and Machines" GE white paper, November 26, 2012.

- Bacidore, Mike. "Are your prepared to work in an autonomous plant?", PlantService, March 2013.

- 

- "Industrial Internet 101 - A Beginner's Guide to the Next Industrial Revolution".

- "NSF Industry/University Cooperative Research Center on Intelligent Maintenance Systems"

# Chapter 22

# Predix (software)

**Predix** is General Electric's software platform for the Industrial Internet.[1]

## 22.1 Industrial application

**Predix** as a cloud-based PaaS (platform as a service) is claimed to enable industrial-scale analytics for asset performance management (APM) and operations optimization by providing a standard way to connect machines, data, and people. GE expects Predix software to do for factories and plants what Apple's iOS did for cell phones.[2] The software was introduced to the market and made available to all companies in 2015.[3] Built on Cloud Foundry open source technology, Predix provides a microservices based delivery model with a distributed architecture (cloud, and on-machine).[4]

## 22.2 See also

- Cloud-based design and manufacturing

- Big data

- SCADA

- Industrial Ethernet

- Internet of Things

- Machine to machine

- Industrial control system

- Industry 4.0

- Intelligent Maintenance Systems

- Cyber-physical system

- Machine-generated data

## 22.3 References

[1] GE. "Predix Powers Industrial-strength Apps". gesoftware.com. Retrieved 2015-06-29.

[2] Passeri, James (2015-03-06). "GE Expects Predix Software to Do for Factories What Apple's iOS Did for Cell Phones". thestreet.com. Retrieved 2015-06-29.

[3] Passeri, James. "GE to Open Up Predix Industrial Internet Platform to All Users". businesswire.com. Retrieved 2015-06-29.

[4] Gee, Sue (24 June 2015). "Predix - A Platform for the Industrial Internet Of Things".

## 22.4 External links

- Official website

# Chapter 23

# Industry 4.0

**Industry 4.0**, **Industrie 4.0** or the **fourth industrial revolution**, is a collective term embracing a number of contemporary automation, data exchange and manufacturing technologies. It had been defined as 'a collective term for technologies and concepts of value chain organization' which draws together Cyber-Physical Systems, the Internet of Things and the Internet of Services.[1][2][3]

Industrie 4.0 facilitates the vision and execution of a "Smart Factory". Within the modular structured Smart Factories of Industry 4.0, cyber-physical systems monitor physical processes, create a virtual copy of the physical world and make decentralized decisions. Over the Internet of Things, cyber-physical systems communicate and cooperate with each other and with humans in real time, and via the Internet of Services, both internal and cross-organizational services are offered and utilized by participants of the value chain.[1]

## 23.1 Name

The term "Industrie 4.0" originates from a project in the high-tech strategy of the German government, which promotes the computerization of manufacturing.[4] The first industrial revolution mobilised the mechanization of production using water and steam power. The second industrial revolution then introduced mass production with the help of electric power, followed by the digital revolution and the use of electronics and IT to further automate production.[5]

The term was first used in 2011 at the Hannover Fair.[6] In October 2012 the Working Group on Industry 4.0 chaired by Siegfried Dais (Robert Bosch GmbH) and Kagermann (acatech) presented a set of Industry 4.0 implementation recommendations to the German federal government. On 8 April 2013 at the Hanover Fair, the final report of the Working Group Industry 4.0 was presented.[7]

## 23.2 Design Principles

There are six design principles in Industry 4.0. These principles support companies in identifying and implementing Industry 4.0 scenarios.[1]

- Interoperability: the ability of cyber-physical systems (i.e. workpiece carriers, assembly stations and products), humans and Smart Factories to connect and communicate with each other via the Internet of Things and the Internet of Services

- Virtualization: a virtual copy of the Smart Factory which is created by linking sensor data (from monitoring physical processes) with virtual plant models and simulation models

- Decentralization: the ability of cyber-physical systems within Smart Factories to make decisions on their own

- Real-Time Capability: the capability to collect and analyse data and provide the derived insights immediately

- Service Orientation: offering of services (of cyber-physical systems, humans or Smart Factories) via the Internet of Services

- Modularity: flexible adaptation of Smart Factories to changing requirements by replacing or expanding individual modules

## 23.3   Meaning

Characteristic for industrial production in an Industry 4.0 environment are the strong customization of products under the conditions of high flexibilized (mass-) production. The required automation technology is improved by the introduction of methods of self-optimization, self-configuration,[8] Self-diagnosis, cognition and intelligent support of workers in their increasingly complex work.[9] The largest project in Industry 4.0 at the present time (July 2013) is the BMBF leading-edge cluster "Intelligent Technical Systems OstWestfalenLippe (it's OWL)". Another major project is the BMBF project RES-COM,[10] as well as the Cluster of Excellence "Integrative Production Technology for High-Wage Countries".[11]

## 23.4   Effects

In June 2013, consultancy firm McKinsey [12] released an interview featuring an expert discussion between executives at Robert Bosch - Siegfried Dais (Partner of the Robert Bosch Industrietreuhand KG) and Heinz Derenbach (CEO of Bosch Software Innovations GmbH) - and McKinsey experts. This interview addressed the prevalence of the Internet of Things in manufacturing and the consequent technology-driven changes which promise to trigger a new industrial revolution. At Bosch, and generally in Germany, this phenomenon is referred to as Industry 4.0. The basic principle of Industry 4.0 is that by connecting machines, work pieces and systems, businesses are creating intelligent networks along the entire value chain that can control each other autonomously.

Some examples for Industry 4.0 are machines which can predict failures and trigger maintenance processes autonomously or self-organized logistics which react to unexpected changes in production.

According to Dais, "it is highly likely that the world of production will become more and more networked until everything is interlinked with everything else". While this sounds like a fair assumption and the driving force behind the Internet of Things, it also means that the complexity of production and supplier networks will grow enormously. Networks and processes have so far been limited to one factory. But in an Industry 4.0 scenario, these boundaries of individual factories will most likely no longer exist. Instead, they will be lifted in order to interconnect multiple factories or even geographical regions.

There are differences between a typical traditional factory and an Industry 4.0 factory. In the current industry environment, providing high-end quality service or product with the least cost is the key to success and industrial factories are trying to achieve as much performance as possible to increase their profit as well as their reputation. In this way, various data sources are available to provide worthwhile information about different aspects of the factory. In this stage, the utilization of data for understanding current operating conditions and detecting faults and failures is an important topic to research. e.g. in production, there are various commercial tools available to provide Overall Equipment Effectiveness (OEE) information to factory management in order to highlight the root causes of problems and possible faults in the system. In contrast, in an Industry 4.0 factory, in addition to condition monitoring and fault diagnosis, components and systems are able to gain self-awareness and self-predictiveness, which will provide management with more insight on the status of the factory. Furthermore, peer-to-peer comparison and fusion of health information from various components provides a precise health prediction in component and system levels and force factory management to trigger required maintenance at the best possible time to reach just-in time maintenance and gain near zero downtime.[13]

## 23.5   Challenges

Challenges which have been identified include

- Lack of adequate skill-sets to expedite the march towards fourth industrial revolution

- Threat of redundancy of the corporate IT department

- General reluctance to change by stakeholders

## 23.6   Role of big data and analytics

Modern information and communication technologies like Cyber-Physical Systems, Big Data or Cloud Computing will help predict the possibility to increase productivity, quality and flexibility within the manufacturing industry and thus to understand advantages within the competition.

Big Data Analytics consists of 6Cs in the integrated Industry 4.0 and Cyber Physical Systems environment. 6C system that is consist of Connection (sensor and networks), Cloud (computing and data on demand), Cyber (model & memory), Content/context (meaning and correlation), Community (sharing & collaboration), and Customization (personalization and value). In this scenario and in order to provide useful insight to the factory management and gain correct content, data has to be processed with advanced tools (analytics and algorithms) to generate meaningful information. Considering the presence of visible and invisible issues in an industrial factory, the information generation algorithm has to be capable of detecting and addressing invisible issues such as machine degradation, component wear, etc in the factory floor.[14][15]

## 23.7   Impact of Industry 4.0

The fourth industrial revolution will affect many areas. Five key impact areas emerge:

1. Machine safety

2. Industry value chain

3. Workers

4. Socio-economic

5. Industry Demonstration: To help industry understand the impact of Industry 4.0, Cincinnati Mayor, John Cranley, signed a proclamation to state "Cincinnati to be Industry 4.0 Demonstration City".[16]

## 23.8   See also

- Big data

- Computer-integrated manufacturing

- Digital modeling and fabrication

- Industrial control system

- Industrial Internet

- Intelligent Maintenance Systems

- Internet of Things

- Machine to machine

- Predictive manufacturing system

- SCADA

## 23.9 References

[1] Hermann, Pentek, Otto, 2015: Design Principles for Industrie 4.0 Scenarios, accessed on 3 February 2015

[2] Jürgen Jasperneite:*Was hinter Begriffen wie Industrie 4.0 steckt* in *Computer & Automation*, 19 Dezember 2012 accessed on 23 December 2012

[3] Kagermann, H., W. Wahlster and J. Helbig, eds., 2013: Recommendations for implementing the strategic initiative Industrie 4.0: Final report of the Industrie 4.0 Working Group

[4] Zukunftsprojekt Industrie 4.0

[5] Die Evolution zur Industrie 4.0 in der Produktion Last download on 14. April 2013

[6] Industrie 4.0: Mit dem Internet der Dinge auf dem Weg zur 4. industriellen Revolution, VDI-Nachrichten, April 2011

[7] Industrie 4.0 Plattform Last download on 15. Juli 2013

[8] Selbstkonfiguierende Automation für Intelligente Technische Systeme, Video, last download on 27. Dezember 2012

[9] Jürgen Jasperneite; Oliver, Niggemann: Intelligente Assistenzsysteme zur Beherrschung der Systemkomplexität in der Automation. In: ATP edition - Automatisierungstechnische Praxis, 9/2012, Oldenbourg Verlag, München, September 2012

[10] Projekt RES-COM

[11] Webseite Exzellenzcluster "Integrative Produktionstechnik für Hochlohnländer", Last download on 15. July 2013

[12] The Internet of Things and the future of manufacturing,

[13] Lee, Jay, Industry 4.0 in Big Data Environment, Harting Tech News 26, 2013, http://www.harting.com/fileadmin/harting/documents/lg/hartingtechnologygroup/news/tec-news/tec-news26/EN_tecNews26.pdf

[14] Lee, Jay; Bagheri, Behrad; Kao, Hung-An (2014). "Recent Advances and Trends of Cyber-Physical Systems and Big Data Analytics in Industrial Informatics". *IEEE Int. Conference on Industrial Informatics (INDIN) 2014.*

[15] Lee, Jay; Lapira, Edzel; Bagheri, Behrad; Kao, Hung-an. "Recent advances and trends in predictive manufacturing systems in big data environment". *Manufacturing Letters* **1** (1): 38–41. doi:10.1016/j.mfglet.2013.09.005.

[16] http://www.imscenter.net/IMS_news/cincinnati-mayor-proclaimed-cincinnati-to-be-industry-4-0-demonstration-city

## 23.10 External links

- Roadmap to the Internet of Things - Its Impact, Architecture & Future Governance - Mark Fell, Carré & Strauss, 2014.

- Cloud-based design and manufacturing

- Industrie 4.0 – Hightech-Strategie der Bundesregierung

- Bundesministerium für Forschung und Entwicklung - Zukunftsprojekt Industrie 4.0

- Plattform Industrie 4.0

- Recommendations for implementing the strategic initiative INDUSTRIE 4.0 English edition - www.plattform-i40.de/

- BMBF-Spitzencluster„Intelligente technische Systeme OstwestfalenLippe it's OWL

- Exzellenzcluster Integrative Produktionstechnik für Hochlohnländer

-

# Chapter 24

# Kaya identity

The **Kaya identity** is an equation relating factors that determine the level of human impact on climate, in the form of emissions of the greenhouse gas carbon dioxide. This identity states that total emission level can be expressed as the product of four inputs: population, GDP per capita, energy use per unit of GDP, carbon emissions per unit of energy consumed.[1] This equation is both very simple and tricky, as it can be reduced to only two terms, but it is developed so that the carbon emission calculation becomes easy, as per the available data, or generally in which format the data is available.

The Kaya identity is a concrete form of the more general I = PAT equation. The latter seeks to describe environmental (I)mpact in terms of the factors (P)opulation, (A)ffluence and (T)echnology. In the Kaya identity, impact is carbon emissions, while technology is split into energy use per unit of GDP and carbon emissions per unit of energy consumed.[2]

## 24.1   Overview

The Kaya identity was developed by Japanese energy economist Yoichi Kaya.[3] It is the subject of his book *Environment, Energy, and Economy: strategies for sustainability* co-authored with Keiichi Yokobori as the output of the *Conference on Global Environment, Energy, and Economic Development (1993 : Tokyo, Japan)*.

Kaya identity is expressed in the form:

$$F = P \times \frac{G}{P} \times \frac{E}{G} \times \frac{F}{E}$$

where:

$F$ is global CO2 emissions from human sources

$P$ is global population

$G$ is world GDP

$E$ is global energy consumption[4]

## 24.2   Use in IPCC reports

The Kaya identity plays a core role in the development of future emissions scenarios in the IPCC Special Report on Emissions Scenarios. The scenarios set out a range of assumed conditions for future development of each of the four inputs. Population growth projections are available independently from demographic research; GDP per capita trends are available from economic statistics and econometrics; similarly for energy intensity and emission levels. The projected carbon emissions can drive carbon cycle and climate models to predict future $CO_2$ concentration and climate change.

## 24.3   Use in other scientific analysis

The Kaya identity is reviewed in a 2002 paper.[5]

A 2007 article[6] uses the Kaya Identity in its analysis of recent trends in carbon emissions, and finds:

> ... cessation or reversal of earlier declining trends in the energy intensity of gross domestic product (GDP) (energy/GDP) and the carbon intensity of energy (emissions/energy), coupled with continuing increases in population and per-capita GDP. Nearly constant or slightly increasing trends in the carbon intensity of energy have been recently observed in both developed and developing regions. No region is decarbonizing its energy supply.

## 24.4   References

[1] Kaya, Yoichi; Yokoburi, Keiichi (1997). *Environment, energy, and economy : strategies for sustainability*. Tokyo [u.a.]: United Nations Univ. Press. ISBN 9280809113.

[2] Nebojsa Nakicenovic and Rob Swart, ed. (2000). "IPCC Special Report on Emissions Scenarios". Retrieved 19 September 2014. |chapter= ignored (help)

[3] Rehmeyer, Julie. "Yoichi Kaya's carbon fix formula". Wired. Retrieved 20 July 2012.

[4] https://www.e-education.psu.edu/meteo469/node/213

[5] Waggoner, P. E.; J. H. Ausubel (2002). "A framework for sustainability science: A renovated IPAT identity" (PDF). *PNAS* **99** (12): 7860–5. Bibcode:2002PNAS...99.7860W. doi:10.1073/pnas.122235999. PMC 122985. PMID 12060732.

[6] Raupach, M.R. et al. (May 22, 2007). "Global and regional drivers of accelerating CO2 emissions" (PDF). *PNAS* **104** (24): 10288–10293. Bibcode:2007PNAS..10410288R. doi:10.1073/pnas.0700609104. PMC 1876160. PMID 17519334.

## 24.5   External links

- Climate change - what is Kaya's equation
- http://www.computare.org/Support%20documents/Fora%20Input/CCC2006/Sustainable%20Paper%2006_05.htm
- http://www.realclimate.org/index.php?p=164
- Online 'Kaya Calculator'

# Chapter 25

# List of emerging technologies

This is a **list of currently emerging technologies**, which contains some of the most prominent ongoing developments, advances, and innovations in various fields of modern technology. Emerging technologies are those technical innovations which represent progressive developments within a field for competitive advantage.[1]

## 25.1   Agriculture

## 25.2   Climate engineering

## 25.3   Construction

### 25.3.1   Architecture

### 25.3.2   Materials science

## 25.4   Displays

## 25.5   Home appliance

## 25.6   Electronics

## 25.7   Energy

## 25.8   Entertainment

## 25.9   IT and communications

## 25.10   Medical

### 25.10.1   Neuroscience

## 25.11   Military

## 25.12   Outer space

## 25.13   Robotics

## 25.14   Transport

See also: List of proposed future transport

## 25.15   See also

**General**  Disruptive innovation, Industrial Ecology, List of inventors, List of inventions, Sustainable development, Technology
readiness level, Anthropogenics

**Ethics** Casuistry, Computer ethics, Engineering ethics, Nanoethics, Bioethics, Neuroethics, Roboethics

## 25.16 Further reading

- Ten Breakthrough Technologies in 2015, *MIT Technology Review*

## 25.17 References

[1] International Congress Innovation and Technology XXI: Strategies and Policies Towards the XXI Century, & Soares, O. D. D. (1997). Innovation and technology: Strategies and policies. Dordrecht: Kluwer Academic.

[2] A review of automation and robotics for the bio-industry. Journal of Biomechatronics Engineering Vol. 1, No. 1, (2008) 37-54

[3] NASA - Investigation of a Closed Ecological System. nasa.gov

[4] Ben Armentrout, and Heidi Kappes. Studies in Closed Ecological Systems: Biosphere in a Bottle

[5] Frieda B. Taub Annual Review of Ecology and Systematics Vol. 5, (1974), pp. 139-160

[6] "Is in vitro meat the future?". The Times. 9 May 2008. Retrieved 7 December 2012.

[7] "Coming soon, the test-tube burger: Lab-grown meat 'needed to feed the world'". Daily Mail. 27 June 2011. Retrieved 18 November 2011.

[8] "Artificial meat: Hamburger junction". The Economist. 25 February 2012. Retrieved 3 March 2012.

[9] "Vertical farming - Does it really stack up?". Te Economist. 9 December 2010. Retrieved 18 November 2011.

[10] "Vertical Farming - Can Urban Agriculture Feed a Hungry World?". Spiegel. 22 July 2011. Retrieved 18 November 2011.

[11] "Vertical Harvest of Jackson Hole Will Grow Organic Produce Even in Winter". Inhabitat. 6 June 2012. Retrieved 29 August 2013.

[12] "3D Printing May Bring Legal Challenges, Group Says". PC World. 10 November 2010. Retrieved 16 April 2011.

[13] "3D printing: the technology that could re-shape the world". The Telegraph. 28 July 2011. Retrieved 18 November 2011.

[14] "3D printer builds houses in China - video". *the Guardian*. 29 April 2014.

[15] "World's First 3D Printed Car Took Years to Design, But Only 44 Hours to Print". *Mashable*. 16 September 2014.

[16] http://www.3ders.org/articles/20140722-neotech-releases-new-system-for-3d-printing-electronics.html

[17] http://3dprint.com/6853/drawn-3d-printed-furniture/

[18] Jelmer Luimstra. "This is a 3D Printer That Can Print Clothes". *3D Printing*.

[19] "World's first climate-controlled domed city to be built in Dubai (PHOTOS)". *rt.com*.

[20] "Sto AG, Cabot Create Aerogel Insulation". Construction Digital. 15 November 2011. Retrieved 18 November 2011.

[21] "Is graphene a miracle material?". BBC Click. 21 May 2011. Retrieved 18 November 2011.

[22] "Could graphene be the new silicon?". The Guardian. 13 November 2011. Retrieved 18 November 2011.

[23] "Applications of Graphene under Development". understandingnano.com.

[24] "The 'new age' of super materials". BBC News. 5 March 2007. Retrieved 27 April 2011.

[25] "Strides in Materials, but No Invisibility Cloak". The New York Times. 8 November 2010. Retrieved 21 April 2011.

[26] NAE Website: Frontiers of Engineering. Nae.edu. Retrieved 22 February 2011.

[27] "Carbon nanotubes used to make batteries from fabrics". BBC News. 21 January 2010. Retrieved 27 April 2011.

[28] "Researchers One Step Closer to Building Synthetic Brain". Daily Tech. 25 April 2011. Retrieved 27 April 2011.

[29] "Pentagon Developing Shape-Shifting 'Transformers' for Battlefield". Fox News. 10 June 2009. Retrieved 26 April 2011.

[30] "Intel: Programmable matter takes shape". ZD Net. 22 August 2008. Retrieved 2 January 2012.

[31] "'Quantum dots' to boost performance of mobile cameras". BBC News. 22 March 2010. Retrieved 16 April 2011.

[32] "3D display technology is no headache, claim researchers". The Engineer. 5 April 2011. Retrieved 17 April 2011.

[33] "'Glasses-free 3D' hits big time as Toshiba sets a date and price for 55-inch set - with a resolution FOUR TIMES hi-def". Daily Mail. 8 December 2011. Retrieved 22 December 2011.

[34] "Scientist: Holographic television to become reality". CNN. 7 October 2008. Retrieved 29 April 2011.

[35] "Holographic video takes step forward with updated display". The Independent. 5 November 2010. Retrieved 29 April 2011.

[36] "Phone calls in 3-D soon". New Straits Times. 9 January 2011. Retrieved 29 April 2011.

[37] "Natural iridescence harnessed for reflective displays". EE Times. 26 July 2009. Retrieved 5 May 2011.

[38] "LG set to sell 55-inch TV using new OLED technology". Taipei Times. 2 January 2012. Retrieved 3 January 2011.

[39] "The TV of the future arrives early: Incredible pictures of 55-inch flatscreen just 4mm thick". Daily Mail. 2 January 2012. Retrieved 3 January 2011.

[40] "LG Announces 4K OLED TVs For Sale". Forbes. 25 August 2014. Retrieved 31 August 2014.

[41] McManamon P.F.; et al. (15 May 1996). "Optical phased array technology". *Proceedings of the IEEE, Laser radar applications* (IEEE) **84** (2): 99–320. Retrieved 29 April 2011.

[42] Wowk B (1996). "Phased Array Optics". In BC Crandall. *Molecular Speculations on Global Abundance*. MIT Press. pp. 147–160. ISBN 0-262-03237-6. Retrieved 29 April 2011.

[43] "Google 'to unveil' hi-tech Google Glasses that put a screen of information over the world". Daily Mail. 20 December 2011. Retrieved 22 December 2011.

[44] "Xeros bead washing machine system set to save water and energy in the home". 19 October 2014.

[45] "QARNOT'S SMART HEATERS DEMOCRATIZE HIGH-POWER COMPUTING". 4 July 2014.

[46] "Vortex-Based Technology Cools Drinks In Less Than A Minute". 21 September 2013.

[47] "Tuberculosis breakthrough as scientists get funds for 'electronic nose'". The Guardian. 7 November 2011. Retrieved 4 December 2011.

[48] "Now, a mobile phone that can smell". The Times of India. 7 November 2011. Retrieved 4 December 2011.

[49] "Electronic Cotton". IEEE Spectrum. January 2012. Retrieved 4 March 2012.

[50] "Flexible future: Forget the iPhone, here's the smartphone made out of 'paper' that will shape with your pocket". Daily Mail. 6 May 2011. Retrieved 18 November 2011.

[51] "Technology... or magic? Samsung shows off video of transparent, flexible screen - with 3D so real it looks like you can touch it". Daily Mail. 7 December 2011. Retrieved 7 December 2011.

[52] "Remapping Computer Circuitry to Avert Impending Bottlenecks". The New York Times. 28 February 2011. Retrieved 27 April 2011.

[53] "Memristor revolution backed by HP". BBC News. 2 September 2010. Retrieved 27 April 2011.

[54] U.S. Patent 7,203,789

[55] U.S. Patent 7,302,513

[56] U.S. Patent 7,359,888

[57] U.S. Patent 7,609,086

[58] U.S. Patent 7,902,857

[59] U.S. Patent 7,902,867

[60] U.S. Patent 8,113,437

[61] "Memristor Models for Pattern Recognition Systems". *springer.com*.

[62] "Scientists developing spintronic computer chips". The Times of India. 14 April 2011. Retrieved 17 April 2011.

[63] "Getting wind farms off the ground". The Economist. 7 June 2007. Retrieved 7 December 2011.

[64] "Wind turbines take to the skies". BBC News. 3 June 2010. Retrieved 7 December 2011.

[65] "Airborne Wind Turbines". The New York Times. 9 December 2007. Retrieved 7 December 2011.

[66] "Solar 'Artificial Leaf' Is Unveiled by Researchers". Reuters. 29 March 2011. Retrieved 21 April 2011.

[67] Faunce TA, Lubitz W, Rutherford AW, MacFarlane D, Moore, GF, Yang P, Nocera DG, Moore TA, Gregory DH, Fukuzumi S, Yoon KB, Armstrong FA, Wasielewski MR, Styring S. Energy and Environment Case for a Global Project on Artificial Photosynthesis. Energy and Environmental Science (2013) DOI: 10.1039/c3ee00063j http://pubs.rsc.org/en/content/articlelanding/2013/ee/c3ee00063j (accessed 2 June 2013)

[68] "Lufthansa to start trials with biofuel from Neste". Reuters. 29 November 2010. Retrieved 4 May 2011.

[69] "California approves Tessera solar plant". Reuters. 28 October 2010. Retrieved 4 May 2011.

[70] "Ultracapacitors Gain Traction as Battery Alternative". Reuters. 30 March 2011. Retrieved 4 May 2011.

[71] "MIT develops way to bank solar energy at home". Reuters. 31 July 2008. Retrieved 24 December 2011.

[72] "All eyes on Bloom Box fuel cell launch". The Guardian. 22 February 2010. Retrieved 24 December 2011.

[73] "Japan aims its home fuel cells at Europe". BBC News. 12 March 2010. Retrieved 24 December 2011.

[74] "Hydrogen tries again". The Economist. 23 April 2010. Retrieved 4 May 2011.

[75] Kraytsberg A, Ein-Eli Y (2011). "Review on Li-air batteries - Opportunities, limitations and perspective". Journal of Power Sources,196:p.886-893.

[76] "Scientists say paper battery could be in the works". Reuters. 7 December 2009. Retrieved 4 May 2011.

[77] "Microfiber fabric makes its own electricity?". Reuters. 14 February 2008. Retrieved 4 May 2011.

[78] "A new device to tap more solar energy invented". The Hindu. 18 May 2011. Retrieved 24 December 2011.

[79] "Heat scavenging - Stealing the heat". The Economist. 4 March 2010. Retrieved 24 December 2011.

[80] "New Rays". Businessworld. 21 May 2011. Retrieved 24 December 2011.

[81] "Electricity to power 'smart grid'". BBC News. 18 May 2009. Retrieved 4 March 2012.

[82] "Smart Grid Costs Are Massive, but Benefits Will Be Larger, Industry Study Says". The New York Times. 25 May 2011. Retrieved 4 March 2012.

[83] "Pushing the low carbon boundaries: South Korea's smart grid initiative". The Guardian. 5 September 2011. Retrieved 4 March 2012.

[84] "Solar panel roads 'could solve energy crisis'". The Telegraph. 8 September 2009. Retrieved 9 December 2011.

[85] "Solar-Powered Glass Road Could Melt Snow Automatically". Fox News. 2 February 2011. Retrieved 9 December 2011.

[86] "Road network could become solar power grid". Wired. 24 September 2010. Retrieved 9 December 2011.

[87] "Wireless energy promise powers up". BBC News. 7 June 2007. Retrieved 4 May 2011.

[88] Parag and Ayesha Khanna. "Do We Need Actors? CGI and the Future of Hollywood". *Big Think*.

[89] IBM *Next Five in Five 2010* (a prediction of five emerging technologies expected in by 2015). See IBM's new 'Next Five in Five' list peers into the future, TechHerald 24 December 2010.

[90] "The big plan to build a brain". The Telegraph. 21 June 2011. Retrieved 18 November 2011.

[91] "IBM's Watson supercomputer crowned Jeopardy king". BBC News. 17 February 2011. Retrieved 17 April 2011.

[92] "MIT scientists take a step closer to artificial intelligence". Computer Weekly. 18 November 2011. Retrieved 18 November 2011.

[93] "How innovative is Apple's new voice assistant, Siri?". New Scientist. 3 November 2011. Retrieved 4 March 2012.

[94] "The People Who Burn Bitcoins". Minyanville. 16 April 2014. Retrieved 20 June 2014.

[95] "G.E.'s breakthrough can put 100 DVDs on a disc". Tehran Times. 28 April 2009. Retrieved 29 April 2011.

[96] "eyeSight CEO Gideon Shmuel: The Company Making Minority Report A Reality (Finally)". The Huffington Post. 30 July 2012. Retrieved 31 July 2012.

[97] "Get Sampark, go multilingual". The Hindu. 3 April 2011. Retrieved 29 April 2011.

[98] "Mobile Video Collaboration System Securely Connects Field Staff and Experts". *Electronic Component News*. 28 March 2011.

[99] "New hand-held device targets work on shop floor: veteran high-tech team launches new venture". *Winnipeg Free Press*. 11 July 2005.

[100] "US scientists build first 'antilaser'". ABC. 18 February 2011. Retrieved 21 April 2011.

[101] "Quantum computing device hints at powerful future". BBC News. 22 March 2011. Retrieved 17 April 2011.

[102] "First Ever Commercial Quantum Computer Now Available for $10 Million". ExtremeTech. 20 May 2011. Retrieved 22 May 2011.

[103] "Does quantum mechanics offer the best way to protect our most valuable data?". The Independent. 31 March 2011. Retrieved 17 April 2011.

[104] "Will NFC make the mobile wallet work?". BBC News. 7 October 2011. Retrieved 8 December 2011.

[105] "Internet of things: Should you worry if your jeans go smart?". BBC News. 23 September 2011. Retrieved 8 December 2011.

[106] "RFID tagging: Chips with everything". The Telegraph. 20 May 2009. Retrieved 8 December 2011.

[107] "Intel goes 3D with transistor redesign". The Guardian. 4 May 2011. Retrieved 19 November 2011.

[108] "Intel unveils 22nm 3D Ivy Bridge processor". BBC News. 4 May 2011. Retrieved 19 November 2011.

[109] "A giant leap into the unknown: GM salmon that grows and grows". The Independent. 22 September 2010. Retrieved 5 May 2011.

[110] "Gene therapy is cure for 'boy in the bubble' syndrome". The Telegraph. 24 August 2011. Retrieved 18 November 2011.

[111] "U.S. Super Soldiers Of The Future Will Be Genetically Modified Transhumans Capable Of Superhuman Feats". *endoftheamericandream.com*.

[112] Joseph Stromberg. "Genetically Modified E. Coli Bacteria Can Now Synthesize Diesel Fuel". *Smithsonian*.

[113] "Genetic Engineering Boosts Ethanol Production by 50 Percent - MIT Technology Review". *MIT Technology Review*.

[114] "USU Synthetic Spider Silk Lab awarded $1.9 million in DOE Energy Efficient Transportation Technology Funds". *CacheValleyDaily.com*.

[115] Shivani Sharma. "Application of Genetic Engineering in Bioremediation: Deinococcus Radiodurans". *biotecharticles.com*.

[116] Katharine Sanderson. "New Portable Kit Detects Arsenic In Wells". *acs.org*.

[117] "Patients to be frozen into state of suspended animation for surgery". The Telegraph. 26 September 2010. Retrieved 21 April 2011.

[118] "Doubt on Anti-Aging Molecule as Drug Trial Stops". The New York Times. 10 January 2011. Retrieved 1 May 2011.

[119] "Signs of ageing halted in the lab". BBC News. 2 November 2011. Retrieved 16 December 2011.

[120] "Stem cells could hold key to 'stopping ageing' say scientists after trial triples mouse lifespan". Daily Mail. 4 January 2012. Retrieved 3 March 2012.

[121] "New Male Birth Control Procedure Is 100 Percent Effective, Completely Reversible [STUDY]". 7 May 2012.

[122] "Researchers see a male counterpart to 'The Pill'". TwinCities.com.

[123] "Aiming for clinical excellence". The Guardian. 26 November 2011. Retrieved 16 December 2011.

[124] "Nanotechnoglogy world: Nanomedicine offers new cures". The Guardian. 6 September 2011. Retrieved 16 December 2011.

[125] "Genetic test could be used to 'personalise' drugs, say scientists". The Independent. 1 March 2010. Retrieved 16 April 2011.

[126] "Scientists Grow Viable Urethras From Boys' Cells". Fox News. 8 March 2011. Retrieved 27 April 2011.

[127] "Doctors grapple with the value of robotic surgery". Houston Chronicle. 16 September 2011. Retrieved 24 December 2011.

[128] "Robotic surgery making inroads in many medical procedures". The Jakarta Post. 8 March 2011. Retrieved 24 December 2011.

[129] "Doctors Perform First Fully Robotic Surgery". PC World. 21 October 2010. Retrieved 24 December 2011.

[130] "Scientists make eye's retina from stem cells". BBC News. 6 April 2011. Retrieved 27 April 2011.

[131] "Medical marvels". The Guardian. 30 January 2009. Retrieved 16 December 2011.

[132] "'Artificial life' breakthrough announced by scientists". BBC News. 20 May 2010. Retrieved 29 April 2011.

[133] "Scientist Craig Venter creates life for first time in laboratory sparking debate about 'playing god'". The Telegraph. 20 May 2010. Retrieved 29 April 2011.

[134] "Artificial blood vessels created on a 3D printer". BBC News. 16 September 2011. Retrieved 26 December 2011.

[135] "Penis tissue replaced in the lab". BBC News. 10 November 2009. Retrieved 26 December 2011.

[136] "U.S. scientists create artificial lungs, of sorts". Reuters. 24 June 2010. Retrieved 26 December 2011.

[137] "Silkworms could aid a breakthrough in tissue engineering". BBC News. 15 December 2011. Retrieved 26 December 2011.

[138] "Research updates: CMU prof wins grants for very cool technology". Pittsburgh Post-Gazette. 11 August 2010. Retrieved 18 November 2011.

[139] "DARPA Program Seeks to Use Brain Implants to Control Mental Illness - MIT Technology Review". MIT Technology Review.

[140] "Mind-reading research: the major breakthroughs". The Telegraph. 22 September 2011. Retrieved 18 November 2011.

[141] "'Mind-reading device' recreates what we see in our heads". The Telegraph. 22 September 2011. Retrieved 18 November 2011.

[142] "It was only a matter of time: Study shows how scientists can now 'read your mind'". Daily Mail. 26 December 2011. Retrieved 27 December 2011.

[143] "Mind-reading device could become reality". The Telegraph. 31 January 2012. Retrieved 4 March 2012.

[144] "IBM predicts it will be making 'mind-controlled' PCs within five years". Daily Mail. 21 December 2011. Retrieved 21 December 2011.

[145] "'Mind Control' Possible in 5 Years: IBM". NBC Bay Area. 21 December 2011. Retrieved 21 December 2011.

[146] Rachel Kaufman (28 January 2011). "New Invisibility Cloak Closer to Working "Magic"". National Geographic News. Retrieved 4 February 2011.

[147] "Breakthrough in bid to create 'invisibility cloak' as 3D object is made to vanish for first time". Daily Mail. 26 January 2012. Retrieved 3 March 2012.

[148] "Laser gun tested on US Navy ship in Pacific Ocean". BBC News. 11 April 2011. Retrieved 18 April 2011.

[149] "Electromagnetic weapons - Frying tonight". The Economist. 15 October 2011. Retrieved 19 November 2011.

[150] "Navy Sets World Record With Incredible, Sci-Fi Weapon". Fox News. 10 December 2010. Retrieved 19 November 2011.

[151] "Star Trek-style force-field armour being developed by military scientists". The Telegraph. 20 March 2010. Retrieved 16 April 2011.

[152] "Navy tests new vehicle-mounted laser weapon". 29 September 2014.

[153] "BAe's anti-gravity research braves X-Files ridicule". The Guardian. 27 March 2000. Retrieved 4 December 2011.

[154] "Boeing joins race to defeat gravity". The Telegraph. 30 July 2002. Retrieved 4 December 2011.

[155] "New Imagery of Asteroid Mission". *NASA*.

[156] "Galileo: What does a more accurate sat-nav system mean?". *BBC News*.

[157] "SoftBank robot to reach US in less than a year". *TechnologyTell*.

[158] "Tiny motors may be big in surgery". BBC News. 20 January 2009. Retrieved 21 April 2011.

[159] Christopher Mims (2009). "Exoskeletons Give New Life to Legs". Scientific American. Retrieved 21 April 2009.

[160] "Riders on a swarm". The Economist. 12 August 2010. Retrieved 21 April 2011.

[161] Sharon Gaudin (2 April 2014). "U.S. Navy to test humanoid robotic firefighters". *Computerworld*.

[162] "Airless Tire Promises Grace Under Pressure for Soldiers". Scientific American. 11 August 2008. Retrieved 6 December 2011.

[163] "New tire models to go without air or oil". The Daily Yomiuri. 6 December 2011. Retrieved 6 December 2011.

[164] "Well, blow me: Airless tyre that never goes flat could put an end to punctures". Daily Mail. 22 January 2012. Retrieved 3 March 2012.

[165] http://ntrs.nasa.gov/archive/nasa/casi.ntrs.nasa.gov/20110015936_2011016932.pdf

[166] "BMW's new driverless car still a decade away". Sydney Morning Herald. 23 November 2011. Retrieved 4 December 2011.

[167] "Driverless car navigates streets". Belfast Telegraph. 20 September 2011. Retrieved 4 December 2011.

[168] "Robotic car developed by Oxford University". BBC News. 10 October 2011. Retrieved 4 December 2011.

[169] "On the performance of electrohydrodynamic propulsion".

[170] "Electrohydrodynamic effect offers promise for efficient propulsion in air".

[171] Scott, William B. (27 November 2006), "Morphing Wings", *Aviation Week & Space Technology*

[172] "FlexSys Inc.: Aerospace". Retrieved 26 April 2011.

[173] Kota, Sridhar; Osborn, Russell; Ervin, Gregory; Maric, Dragan; Flick, Peter; Paul, Donald. "Mission Adaptive Compliant Wing – Design, Fabrication and Flight Test" (PDF). Ann Arbor, MI; Dayton, OH, U.S.A.: FlexSys Inc., Air Force Research Laboratory. Retrieved 26 April 2011.

[174] "Showcase UAV Demonstrates Flapless Flight". BAE Systems. 2010. Retrieved 22 December 2010.

[175] "Demon UAV jets into history by flying without flaps". *Metro.co.uk* (London: Associated Newspapers Limited). 28 September 2010. Retrieved 29 September 2010.

[176] "Terrafugia Transition flying car to go into production after US approval". The Australian. 1 July 2010. Retrieved 6 December 2011.

[177] "Stuck in traffic? Turn your car into a plane in 30 secs". The Times of India. 2 July 2010. Retrieved 6 December 2011.

[178] "Rocket powered by nuclear fusion could send humans to Mars". 4 April 2013.

[179] "Flying train unveiled by Japanese scientists". CNN. 13 May 2011. Retrieved 7 December 2011.

[180] "Robot plane-train uses ground-effect principle to levitate". Wired. 12 May 2011. Retrieved 7 December 2011.

[181] "Going up in the world? Beat the rush hour with first commercial jetpack for £50,000". Daily Mail. 26 February 2010. Retrieved 6 December 2011.

[182] "Flying into the future: New Zealand company to make personal jet packs". The Telegraph. 24 February 2010. Retrieved 6 December 2011.

[183] "China scientists claim 1,200 kph train". CNN. 1 February 2011. Retrieved 7 December 2011.

[184] "Laboratory working on train to run at 1,000 kph". Shanghai Daily. 3 August 2010. Retrieved 7 December 2011.

[185] "How vactrains work". The Seattle Times. 16 August 2010. Retrieved 7 December 2011.

[186] "The Potential for Maglev Applications". about.com.

[187] "Pod Cars Start to Gain Traction in Some Cities". The New York Times. 20 September 2010. Retrieved 7 December 2011.

[188] "Are driverless pods the future?". BBC News. 18 December 2007. Retrieved 7 December 2011.

[189] Physical Internet would increase profits, reduce carbon emissions, study finds. Phys.org. Retrieved on 2013-07-21.

[190] "China Developing Scramjet Propulsion". Aviation Week. 2 September 2007. Retrieved 6 December 2011.

[191] "X-51A Scramjet Fails On Second Attempt". Aviation Week. 15 June 2011. Retrieved 6 December 2011.

[192] "On hybrid, space travel will not cost the earth". The Times of India. 23 March 2010. Retrieved 6 December 2011.

[193] "Robot passes test in space elevator contest". The Washington Post. 5 November 2009. Retrieved 19 November 2011.

[194] "Plans to develop space-plane are go". The Australian. 20 October 2008. Retrieved 7 December 2011.

[195] "UK Skylon spaceplane passes key review". BBC News. 24 May 2011. Retrieved 7 December 2011.

[196] "Air Force says it's extending mission of mysterious X-37B". Los Angeles Times. 29 November 2011. Retrieved 7 December 2011.

http://www.techinfopoint.com

# Chapter 26

# Non-Nuclear Futures

**Non-Nuclear Futures: The Case for an Ethical Energy Strategy** is a 1975 book by Amory B. Lovins and John H. Price.[1][2] The main theme of the book is that the most important parts of the nuclear power debate are not technical disputes but relate to personal values, and are the legitimate province of every citizen, whether technically trained or not. Lovins and Price suggest that the personal values that make a high-energy society work are all too apparent, and that the values associated with an alternate view relate to thrift, simplicity, diversity, neighbourliness, craftsmanship, and humility.[3] They also argue that large nuclear generators could not be mass-produced. Their centralization requires costly transmission and distribution systems. They are inefficient, not recycling excess thermal energy. They are much less reliable and take longer to build, exposing them to escalated interest costs, mistimed demand forecasts, and wage pressure by unions.

Lovins and Price suggest that these two different sets of personal values and technological attributes lead to two very different policy paths relating to future energy supplies. The first is high-energy nuclear, centralized, electric; the second is lower energy, non-nuclear, decentralized, less electrified, softer technology.[4]

Subsequent publications by other authors which relate to the issue of non-nuclear energy paths are Greenhouse Solutions with Sustainable Energy, Plan B 2.0, Reaction Time, State of the World 2008, The Clean Tech Revolution, and the work of Benjamin K. Sovacool.

## 26.1   See also

- Anti-nuclear movement in the United States

- Contesting the Future of Nuclear Power

- List of books about nuclear issues

- Nuclear energy policy

- Nuclear or Not?

- Nuclear-Free Future Award

- Nuclear-free zone

- Rocky Mountain Institute

- Kristin Shrader-Frechette

- Benjamin K. Sovacool.

## 26.2 References

[1] Lovins, Amory B. and Price, John H. (1975). *Non-nuclear Futures: The Case for an Ethical Energy Strategy* (Cambridge, Mass.: Ballinger Publishing Company, 1975. xxxii + 223pp. ISBN 0-88410-602-0, ISBN 0-88410-603-9).

[2] Weinberg, Alvin M. (December 1976). "Book review. Non-nuclear futures: the case for an ethical energy strategy". *Energy Policy* (Elsevier Science Ltd.) **4** (4): 363–366. doi:10.1016/0301-4215(76)90031-8. ISSN 0301-4215.

[3] *Non-Nuclear Futures*, pp. xix-xxi.

[4] *Non-Nuclear Futures*, p. xxiii.

## 26.3 External links

- Nuclear Power's Global Expansion: Weighing Its Costs and Risks

# Chapter 27

# Physics of the Future

***Physics of the Future: How Science Will Shape Human Destiny and Our Daily Lives by the Year 2100*** is a 2011 book by theoretical physicist Michio Kaku, author of *Hyperspace* and *Physics of the Impossible*.[1][2] It speculates on the possibilities of future technological development over the next 100 years. Interviewing notable scientists of their field of research Kaku lays out his vision of coming developments in medicine, computing, artificial intelligence, nanotechnology, and energy production,[3] stating that "this book most closely resembles my book *Visions*."[1] Kaku writes how he hopes his predictions for 2100 will be as successful as science fiction writer Jules Verne's 1863 novel *Paris in the Twentieth Century*. Kaku contrasts Verne's foresight against U.S. Postmaster General John Wanamaker, who in 1893 predicted that mail would still be delivered by stagecoach and horseback in 100 years' time, and IBM chairman Thomas J. Watson, who in 1943 is alleged to have said "I think there is a world market for maybe five computers."[4] Kaku points to this long history of failed predictions against progress to underscore his notion "that it is very dangerous to bet against the future".[1] The book was on the New York Times Bestseller List for 5 weeks.[5]

## 27.1 Contents

Each chapter is sorted into three sections: Near future (2000-2030), Midcentury (2030-2070), and Far future (2070-2100). Kaku notes that the time periods are only rough approximations, but show the general time frame for the various trends in the book.[1]

### 27.1.1 Future of the Computer: Mind over Matter

Kaku begins with Moore's law, and compares a chip that sings "Happy Birthday" with the Allied forces in 1945, stating that the chip contains much more power,[1][6] and that "Hitler, Churchill, or Roosevelt might have killed to get that chip." He predicts that computer power will increase to the point where computers, like electricity, paper, and water, "disappear into the fabric of our lives, and computer chips will be planted in the walls of buildings."

He also predicts that glasses and contact lenses will be connected to the internet, using similar technology to virtual retinal displays. Cars will become driverless due to the power of the GPS system. This prediction is supported by the results of the Urban Challenge. The Pentagon hopes to make $1/3$ of the United States ground forces automated by 2015.[1] Technology similar to BrainGate will eventually allow humans to control computers with tiny brain sensors, and "like a magician, move objects around with the power of our minds."

### 27.1.2 Future of AI: Rise of the Machines

Kaku discusses robotic body parts, modular robots, unemployment caused by robots, surrogates and avatars (like their respective movies), and reverse engineering the brain. Kaku goes over the three laws of robotics and their contradictions. He endorses a "chip in robot brains to automatically shut them off if they have murderous thoughts", and believes that the most likely scenario is one in which robots are free to wreak havoc and destruction, but are designed to desire benevolence.[1]

### 27.1.3 Future of Medicine: Perfection and Beyond

Kaku believes that in the future, reprogramming one's genes can be done by using a specially programmed virus, which can activate genes that slow the aging process. Nanotech sensors in a room will check for various diseases and cancer, nanobots will be able to inject drugs into individual cells when diseases are found, and advancements in extracting stem cells will be manifest in the art of growing new organs. The idea of resurrecting an extinct species might now be biologically possible.

### 27.1.4 Nanotechnology: Everything from Nothing?

Kaku discusses programmable matter, quantum computers, carbon nanotubes, and the possibility of replicators. He also expects a variety of nanodevices that search and destroy cancer cells cleanly, leaving normal cells intact.

### 27.1.5 Future of Energy: Energy from the Stars

Kaku discusses the draining of oil on the planet by pointing to the Hubbert curve, and the rising problem of immigrants who wish to live the American dream of wasteful energy consumption. He predicts that hydrogen and solar energy will be the future, noting how Henry Ford and Thomas Edison bet on whether oil or electricity would dominate, and describing fusion with lasers or magnetic fields, and dismisses cold fusion as "a dead end". Kaku suggests that nations are reluctant to deal with global warming because of the extravagance of oil, being the cheapest source of energy, encourages economic growth. Kaku believes that in the far future, room-temperature superconductors will usher the era of magnet-powered floating cars and trains.

### 27.1.6 Future of Space Travel: To the Stars

Unlike conventional chemical rockets which use Newton's third law of motion, solar sails take advantage of radiation pressure from stars. Kaku believes that after sending a gigantic solar sail into orbit, one could install lasers on the moon, which would hit the sail and give it extra momentum.

Another alternative is to send thousands of nanoships, of which only a few would reach their destination. "Once arriving on a nearby moon, they could create a factory to make unlimited copies of themselves," says Kaku. Nanoships would require very little fuel to accelerate. They could visit the stellar neighborhood by floating on the magnetic fields of other planets.

### 27.1.7 Future of Wealth: Winners and Losers

Kaku discusses how Moore's law robotics will affect the future capitalism, which nations will survive and grow, how the United States is "brain-draining" off of immigrants to fuel their economy.

### 27.1.8 Future of Humanity: Planetary Civilization

Kaku ranks the civilization of the future, with classifications based on energy consumption, entropy, and information processing.

## 27.2   Reception

The *Wall Street Journal* considers it a "largely optimistic view of the future".[7] In April 2011, *The Telegraph* stated "[Physics of the Future] is partisan about technology in a way that smacks of Gerard K. O'Neill's deliriously technocratic vision of space exploration, *The High Frontier*."[8] *Kirkus Reviews* stated "The author's scientific expertise will engage readers too sophisticated for predictions based on psychic powers or astrology."[9] Reviewers at *Library Journal* have stated, "This work is highly recommended for fans of Kaku's previous books and for readers interested in science and robotics."[10] Barnes & Noble stated, "Physics of the Future qualifies as one of the most exciting science books of the new millennium."[2] *The Economist* is skeptical about prediction in general pointing out that unforeseen "unknown unknowns" led to many disruptive technologies over the century just past.[11]

Not all reviewers had a positive response to the book. Writing in journal *Physics Today*, physicist Neil Gershenfeld said that the book has "an appealing premise" but describes "a kind of future by committee" populated by "science-fiction staples". Gershenfeld said, "Such a forecast could have been accomplished with less effort by collating covers from popular science magazines." Gershenfeld criticizes Kaku for "some surprising physics errors", such as ignoring air friction on maglev vehicles. Kaku is praised for raising "profound questions", such as the effect of affluence in the future, or the decoupling of sensory experience from reality. However, Gershenfeld laments that these questions are asked in the margins and not given a deep treatment. "It would have been more relevant to learn the author's perspective on these questions than to find out where and to whom he's presented lectures," Gershenfeld said.[12]

## 27.3   References

[1]  Kaku, Michio (March 2011). *Physics of the Future: How Science Will Shape Human Destiny And Our Daily Lives by the Year 2100*. Doubleday. ISBN 978-0-385-53080-4.

[2]  "Physics of the Future, Michio Kaku, (9780385530804). Hardcover - Barnes & Noble". Retrieved 24 May 2011.

[3]  "Physics of the Future: How Science Will Shape Human Destiny and Our Daily Lives by the Year 2100 - Michio Kaku - Hardcover (ISBN 9780385530804)". *Borders Group*. Retrieved 24 May 2011.

[4]  Kaku, Michio. "Physics of the Future". *Excerpt*. NPR. Retrieved 22 June 2011.

[5]  Schuessler, Jennifer (1 May 2011). "Best Sellers". *The New York Times*. Retrieved 24 May 2011.

[6]  Kaku, Michio (March 2011). *Physics of the Future: How Science Will Shape Human Destiny And Our Daily Lives by the Year 2100*. Doubleday. p. 21. ISBN 978-0-385-53080-4.

[7]  "Book Review: Physics of the Future - WSJ.com". *Wall Street Journal*. 23 March 2011. Retrieved 24 May 2011.

[8]  Ings, Simon (26 April 2011). "Physics of the Future by Michio Kaku: review - Telegraph". *The Daily Telegraph* (London). Retrieved 24 May 2011.

[9]  "PHYSICS OF THE FUTURE by Michio Kaku". *Kirkus Reviews*. Retrieved 24 May 2011.

[10]  "Science & Technology Reviews, February 1, 2011". Retrieved 24 May 2011.

[11]  "Suspension of disbelief: known unknowns and unknown unknowns". *The Economist*. March 10, 2011. Retrieved 26 May 2011.

[12]  Gershenfeld, Neil (October 2011). "Physics of the Future". *Physics Today* **64** (10): 56. doi:10.1063/pt.3.1299.

## 27.4   External links

- Michio Kaku's official website

# Chapter 28

# Project Hieroglyph

**Project Hieroglyph** is an initiative to create science fiction that will spur innovation in science and technology founded by Neal Stephenson in 2011.[1]

Stephenson framed the ideas behind Hieroglyph in a World Policy Institute article entitled "Innovation Starvation" [2] where he attempts to rally writers to infuse science fiction with optimism that could inspire a new generation to, as he puts it, "get big stuff done."

Stephenson says that "a good SF universe has a coherence and internal logic that makes sense to scientists and engineers. Examples include Isaac Asimov's robots, Robert Heinlein's rocket ships, and William Gibson's cyberspace. Such icons serve as hieroglyphs—simple, recognizable symbols on whose significance everyone agrees."[3]

Stephenson partnered with Arizona State University's Center for Science and the Imagination[4] which now administers the project.

In September 2014, the project's first book, Hieroglyph: Stories and Visions for a Better Future, edited by Ed Finn and Kathryn Cramer was published by William Morrow. Contributors to the book include Neal Stephenson, Bruce Sterling, Madeline Ashby, Gregory Benford, Rudy Rucker, Vandana Singh, Cory Doctorow, Elizabeth Bear, Karl Schroeder, James Cambias, Brenda Cooper, Charlie Jane Anders, Kathleen Ann Goonan, Lee Konstantinou, Annalee Newitz, Geoffrey Landis, David Brin, Lawrence Krauss, and Paul Davies.

## 28.1 See also

- Collaborative innovation network
- Exploratory engineering
- Fictional technology
- Invention
- Macro-engineering
- Megaproject
- Megascale engineering
- The Mongoliad
- Retrofuturism
- Techno-progressivism
- Technological utopianism

## 28.2   References

[1] "Dear Science Fiction Writers: Stop Being So Pessimistic! | Science & Nature | Smithsonian Magazine". Smithsonianmag.com. Retrieved 2012-04-14.

[2] "Innovation Starvation | World Policy Institute". Worldpolicy.org. Retrieved 2012-04-14.

[3] "Hieroglyph | Home". Hieroglyph.asu.edu. Retrieved 2012-04-14.

[4] "Center for Science and the Imagination, Arizona State University".

## 28.3   Further reading

- Innovation stagnation is slowing U.S. progress by David Brooks, *Houston Chronicle*, October 7, 2011.

- Project Hieroglyph: Fighting society's dystopian future by Debbie Siegelbaum, BBC News, Washington, September 3, 2014.

# Chapter 29

# Project on Emerging Nanotechnologies

*Project Logo*

The **Project on Emerging Nanotechnologies** was established in 2005 as a partnership between the Woodrow Wilson International Center for Scholars and the Pew Charitable Trusts.[1] The Project was intended to address the social, political, and public safety aspects of nanotechnology. It intended in particular to look for research and policy gaps and opportunities in knowledge and regulatory processes, and to develop strategies for closing them. The project worked with multiple U.S. and foreign governments and organizations.

The project's stated goal was "to inform the debate and to create an active public and policy dialogue. It was not an advocate either for, or against, particular nanotechnologies. Rather, the Project sought to ensure that as these technologies are developed, potential human health and environmental risks were anticipated, properly understood, and effectively managed."[2]

## 29.1 Publications

They have produced many publications on the various aspects of nanotechnology policy. One of the notable reports is on *Managing the Effects of Nanotechnology*, written by J. Clarence (Terry) Davies in 2006.[3] They also maintain several online databases including the widely cited consumer products inventory, the *Nanotechnology Health and Environmental Implications: An inventory of current research*[4][5][6] as well as a series of PEN Reports. Their work has also been published in academic journals such as Nature Nanotechnology.[7]

A major activity of the Project was testimony on public forums.[8][9][7]

## 29.2  Staff

- David Rejeski, director, also the Director of the Foresight and Governance Project at the Woodrow Wilson Center, an initiative designed to facilitate long-term thinking and planning in the public sector.

- Todd Kuiken, Policy Associate

- Eleonore Pauwels, Visiting Scholar

The Advisory Board included Linda J. Fisher, Vice President and Chief sustainability officer at DuPont, Margaret A. Hamburg M.D., Vice President for Biological Programs, Nuclear Threat Initiative, Donald Kennedy, editor-in-chief of Science magazine and president emeritus and Bing Professor of Environmental Science, Emeritus, at Stanford University, John Ryan is Director of the Bionanotechnology IRC at Oxford University, and Stan Williams, Senior Fellow and Director of Quantum Science Research at Hewlett-Packard.

## 29.3  References

[1] nanotechproject.org – Mission

[2] About The Project on Emerging Nanotechnologies

[3] Canadian Embassy Science and Technology News, Jan-Feb 2006

[4] nanotechproject.org

[5] "Southern Compass", Southern Growth Policies Board, 2005.

[6] "Nanocosmetics Alarm Safety Advocates" by Michelle Chen, **The New Standard**, Oct.12, 2006.

[7] "Former White House science advisor warns that nanotechnology's potential threatened"

[8] "Environmental and Safety Impacts of Nanotechnology: What Research is Needed?" Testimony of David Rejeski, US House of Representatives, Nov 17, 2005

[9] Hearing on:"Developments in Nanotechnology" Testimony of: Dr. J. Clarence (Terry) Davies, Senior Advisor, Project on Emerging Nanotechnologies U.S. House of Representatives Committee on Commerce, Science and Transportation, Feb. 15, 2006

## 29.4  External links

- Project on Emerging Nanotechnologies

- Woodrow Wilson International Center for Scholars

- The Pew Charitable Trusts

# Chapter 30

# Pure fusion weapon

A **pure fusion weapon** is a hypothetical hydrogen bomb design that does not need a fission "primary" explosive to ignite the fusion of deuterium and tritium, two heavy isotopes of hydrogen (see thermonuclear weapon for more information about fission-fusion weapons). Such a weapon would require no fissile material and would therefore be much easier to build in secret than existing weapons. The necessity of separating high-quality fissile material requires a substantial industrial investment, and blocking the sale and transfer of the needed machinery has been the primary mechanism to control nuclear proliferation to date.

All current thermonuclear weapons use a fission bomb as a first stage to create the high temperatures and pressures necessary to start a fusion reaction between deuterium and tritium in a second stage. For many years, nuclear weapon designers have researched whether it is possible to create high enough temperatures and pressures inside a confined space to ignite a fusion reaction, without using fission. Pure fusion weapons offer the possibility of generating very small nuclear yields and the advantage of reduced collateral damage stemming from fallout because these weapons would not create the highly radioactive byproducts associated with fission-type weapons. These weapons would be lethal not only because of their explosive force, which could be large compared to bombs based on chemical explosives, but also because of the neutrons they generate.

While various Neutron source devices have been developed, some of them based on fusion reactions, none of them are able to produce an energy yield, neither in controlled form for energy production nor uncontrolled for a weapon.

Despite the many millions of dollars spent by the U.S. between 1952 and 1992 to produce a pure fusion weapon, no measurable success was ever achieved. In 1998, the U.S. Department of Energy (DOE) released a restricted data declassification decision stating that even if the DOE made a substantial investment in the past to develop a pure fusion weapon, "*the U.S. Is not known to have and is not developing a pure fusion weapon and no credible design for a pure fusion weapon resulted from the DOE investment*". The power densities needed to ignite a fusion reaction still seem attainable only with the aid of a fission explosion, or with large apparatus such as powerful lasers like those at the National Ignition Facility, the Sandia Z-pinch machine, or various magnetic tokamaks. Regardless of any claimed advantages of pure fusion weapons, building those weapons does not appear to be feasible using currently available technologies and many have expressed concern that pure fusion weapons research and development would subvert the intent of the Nuclear Non-Proliferation Treaty and the Comprehensive Test Ban Treaty.

It has been claimed that it is possible to conceive of a crude, deliverable, pure fusion weapon, using only current day, unclassified technology. The weapon design[1] weighs approximately 3 tonnes, and might have a total yield of approximately 3 tonnes of TNT. The proposed design uses a large explosively pumped flux compression generator to produce the high power density required to ignite the fusion fuel. From the point of view of explosive damage, such a weapon would have no clear advantages over a conventional explosive, but the massive neutron flux could deliver a lethal dose of radiation to humans within a 500 meter radius (most of those fatalities would occur over a period of months, rather than immediately).

Some researchers have examined the use of antimatter as an alternative fusion trigger, mainly in the context of antimatter-catalyzed nuclear pulse propulsion.[2] Such a system, in a weapons context, would have many of the desired properties of a pure fusion weapon. However, the technical barriers to producing and containing the required quantities of antimatter appear formidable, well beyond present capabilities. Induced gamma emission is another approach that is currently being researched. Very high energy-density chemicals such as the mythical red mercury, various ballotechnics and others have also been suggested as a means of triggering a pure fusion weapon.

## 30.1  References

[1] Jones, S. L.; von Hippel, F. N. (1998). "The Question of Pure Fusion Explosions under the CTBT" (pdf). *Science and Global Security* **7**: 129–150. doi:10.1080/08929889808426452.

[2] Antimatter weapons

- Department of Energy, Office of Declassification (1 January 2001). "Restricted Data Declassification Decisions 1946 to the Present". FAS. RDD-7. References to pure fusion weapon are in section V. C. 1. g.

## 30.2  External links

- "Opening Pandora's nuclear war chest", article on "fourth generation" weapons

# Chapter 31

# Real-time Delphi

**Real-time Delphi** (RTD) is an advanced form of the Delphi method. The advanced method "is a consultative process that uses computer technology" [1] to increase efficiency of the Delphi process.

## 31.1   Definition and idea

Gordon and Pease [2] define the advanced approach as an innovative way to conduct Delphi studies that do not involve sequential "rounds" and consequently lead to a higher degree of efficiency with regard to the time frame needed to perform such studies. Friedewald, von Oertzen, and Cuhls [3] underline that aspect by writing, in "a Real-Time-Delphi, the participants do not only judge twice but can change their opinion as often as they like when they see the aggregated results of the other participants". So, here it becomes clear that the Real-Time Delphi approach requires real-time calculation and provision of group responses. Friedewald et al.[3] further state that the Real-Time Delphi method has beneath its explorative and predictive elements also normative and communicative elements. These latter are investigated by Bolognini,[4] who explores the potential of computer-based Delphi as a communication technique for electronic democracy.

## 31.2   History

The basic idea of a real-time, therefore computer-based (usually web-based), Delphi approach originates in a paper published by Turoff back in 1972 about an online Delphi conference conducted in the United States.[5] The conference was characterized by remote locations of participants, an online tool to access and give judgments, anonymity of the participants, continuous operations and analysis of results (i.e. participants were able to see given answers of the other participants in real-time), as well as asynchronous participation (i.e. participants could independently login and logout how often and when they desired). The stated aspects are some of the key characteristics of Real-Time Delphi studies, which shows that the original idea of conducting such studies can be traced back to the respective year. Today, nevertheless, technological innovations and advanced computer aided design possibilities (e.g. high-speed internet connections, high definition graphic, and advanced processor performance) facilitate more sophisticated studies in this context.[6] The general idea to develop a faster advanced form of Delphi studies by using ideas and basic concepts of Turoff, was initiated by the U.S. Defense Advanced Research Projects Agency (DARPA), which awarded a grant in 2004 to develop an approach to improve "speed and efficiency of collecting judgments in tactical situations".[2] A small software company named Articulate Software in San Francisco was awarded an innovation research grant to develop what DARPA was asking for.[2] Adam Pease, principal consultant and CEO of Articulate Software, published the findings and methodology together with Theodore Gordon in 2006.[2]

## 31.3    Differences between Conventional and Real-time Delphi Method

The question arises how a Real-Time Delphi study differs from a Conventional Delphi study. The basic framework is to think of a Delphi study which is conducted in form of an online questionnaire. However, a Conventional round-based Delphi study conducted via the internet is called "Internet Delphi". The basic difference to Internet Delphi is that the process of a Real-Time Delphi is not characterised by single iterated rounds. In fact, real-time calculation and provision of responses are the key characteristics of Real-Time Delphis. Various other labels for Real-Time Delphi can be found in literature and many authors are not completely aware of the differences:[7] "Electronic Delphi", "Computer Delphi", "Computer-aided Delphi", and "Technology Delphi". However, it is important to truly understand the design and process a researcher has chosen to find out whether real-time calculations and provisions have been applied or not.

The typical Real-Time Delphi process can be described in the way that participants get access to an online questionnaire portal for a certain time frame, within which they are allowed to login and logout as often as they want. Whenever they login, they will see all their quantitative and qualitative answers of previous sessions and they can change all answers as desired within the given period of time. Besides their own answers they will see the on-going – hence, real-time – responses of other participants, and with regard to metric assessments the group as a whole will be visualised in terms of median, average, and interquartile range (IQR). It has to be pointed out that the numerical visualisations as well as the qualitative inputs change in the course of other participants changing their responses.[2] Consequently, a participant can find out to what extent his own responses from an earlier point of time are still within the group opinion (i.e. IQR). The core innovation, then, of Real-Time Delphi studies is the real-time calculation and provision of results.[8]

## 31.4    Methodological Advancements

The core methodological innovation of Real-Time Delphi studies are the absence of iterated rounds and the real-time calculation and provision of group responses. Whereas Conventional Delphi studies are characterised by repeating sequential rounds, the Real-Time Delphi approach is characterised by a continuous round-less procedure leading to a reduced time frame needed to conduct such studies.[9] Consequently, conducting large-scale studies of huge complexity in a relatively short period of time becomes possible.[9] Another core methodological innovation is the fact that experts may not only judge once or twice, depending on the number of rounds, as it has been usual in a Conventional Delphi study. During a Real-Time Delphi, experts can independently reassess their responses as often as they want.[3]

Hartman and Baldwin [1] discuss further advantages of the Real-Time Delphi approach: First, the number of experts participating in the real-time study can be increased due to a higher degree of automation during and improved possibilities for analysis after the study. Additionally, the Internet provides the possibility to invite a worldwide expert panel to participate in the study. Second, the degree of interaction among the experts can be increased due to the fact that they can immediately react on others' comments. Additionally, the time frame between giving own answers and getting insights into others' responses is very short, which encourages stronger cognitive examination with the respective issue in question. Hartman and Baldwin [1] argue that with the help of this procedure the validity of results is maximised.

In order to conduct a Real-Time Delphi study, computer software – usually web-based – is needed for facilitating real-time calculations and visualisation of results. It is generally proposed in the existing literature that the experts participating in the study see not only their own answers but also the median and interquartile range of all given responses immediately after answering a quantitative question.[2] Besides the quantitative assessment a qualitative judgment of participants can be shown which serves as a justification for their numerical assessment of the question. Additionally, it can be shown to the expert how many respondents have already given their answers. To examine the qualitative arguments of others participants can click on a button, and a "reasons window" opens, which shows the statements of other participants to underline their point of view. So, the respective legitimisations given by others may cause a respondent to recapture his own point of view.[2] In the next step the expert can change his own answer, add new arguments to underline his point of view, or leave his answer unchanged. In addition, the respondent will be shown an attention indicator, a so-called "flag",[9] if his answer is within or outside the interquartile range or significantly different from the median. This application helps to see and understand immediately the own assessment and to think about reasons for the deviation from group opinions or else a high degree of consent. The respondent's attention will be called by highlighting questions with a high degree of deviation with a different colour and by asking him to give further reasons for his deviation from the group opinion.

After operating a question in the described procedure, the participant can continue to the next question or press a "save"-button in the program, which leads to an immediate update of the median, interquartile range, and given arguments, and then leave the program. A second advantage of the round-less approach is the fact that, in order to take part in the study, participants can login and logout with their personalised account as often as they want during the time frame provided. Their already given answers will be saved and recalled when they login the next time. So, by design of the study, there are no explicit single rounds to answer the questions.[2] Updating and playing back the information to the other participants follow immediately in succession to the process of answering. Here, it becomes clear that the process of answering can be synchronous or asynchronous and a worldwide expert panel can be reached, which is one of the major advantages of web-based tools. Turoff and Hiltz [7] argue that the issue of asynchronous interaction is probably one of the least understood characteristics of Real-Time Delphis. Zipfinger [10] points out two advantages of asynchronous participation of the experts: First, they can login to the portal whenever they want; therefore, one could argue that the degree of convenience of taking part is increased for the participant due to a 24h-availability of the portal. Second, panellists can contribute to whatever aspects in the questionnaire they want, especially when having gone through each question at least once.[7] Here, a substantial aspect of Real-Time Delphis becomes obvious: Turoff and Hiltz [7] explain that a Real-Time Delphi study offers a design of structured communication which allows every individual to choose the sequence and speed to contribute to the problem solution process. So, in comparison to face-to-face discussions, the Real-Time Delphi approach gives room for individuality and different cognitive abilities of the participants.[7]

A further advantage is the fact that the administrator of the study can set an arbitrary time frame in which participants have to login and take part in the questionnaire. So, whenever the researcher or administrator of the study is satisfied with the existing answers (i.e. in terms of quantity and quality), he can declare the study to be ended and close the online tool (i.e. "freeze" the responses).[2][9]

It is obvious that the key features of Delphi studies are also met in the context of Real-Time Delphi studies. Anonymity, controlled feedback, and statistical group response are met as explained in Chapter 2.[2] However, the issue of iteration is, by design of the technique, not valid for Real-Time Delphi studies anymore. Instead of answering each question a first time and getting a second sheet with the group responses in the second round, the Real-Time Delphi already shows the second screen (i.e. group responses) immediately after answering each question. Having answered each question at least once, the participant can usually control which question to reassess from a "consensus portal", which serves as a kind of control panel to access single questions again. So, on the one hand, the procedure differs from a Conventional Delphi and, on the other hand, the iteration into single rounds is missing.

Having asked the question how the accomplishment of a Real-Time Delphi study differs from conducting a usual Delphi study, Gordon and Pease [2] point out that a Real-Time Delphi study can be implemented via a site on the Internet or in any other network (e.g. intra-company network, local area network) and is, therefore, not conducted in paper-and-pencil form any more.

As with all Delphi studies, the process of defining and selecting experts is still extremely important.[2] The Conventional Delphi study is then divided into several steps of response round, analysis through the facilitator, playing back the information, next response round, and so on. However, the Real-Time Delphi study is, after granting access to the online tool, rather a self-running process.

The basic strengths of a Real-Time Delphi study are its efficiency and applicability to all Delphi topics (i.e. common problem sets, decision making issues, cross impact studies, etc.).[2] Figure 2 illustrates that the process of a Real-Time Delphi differs. Important is to point out that the number of interventions of the facilitator needed during the response phases (i.e. after opening the online tool) are usually less. Having developed the online tool in advance, the intermediate analysis done by the facilitator of the study is rather uncomplicated in comparison to the Conventional Delphi. The overall shortened time period needed to conduct a real-time study underlines that the approach can be regarded as generally more efficient.[2]

Gordon and Pease [2] point out that a Real-Time Delphi study is applicable for a wide range of possible circumstances under which the consultation of experts is necessary. On the one hand, the authors give the example of a "small group operating synchronously in a conference room with laptop computers connected wirelessly to the web site where the software resides, with anticipated completion of the exercise in say 20 min.". On the other hand, it can be thought of a larger panel of experts operating asynchronously from remote locations within a longer period of time.

The greatest weakness of the real-time approach is that it is missing a wholly integrated, scientifically founded concept. The real-time Delphi idea is still a very new concept, which requires further research and application to become a tool for full scale operations.[2] Especially the editing of the alpha (i.e. the first) inputs of respondents, the real-time presentation of group results, and the tracking of progress over time should be integrated in a kind of administrator package to make the accomplishment of a Real-Time Delphi less difficult.[2]

## 31.5   Examples of Real-time Delphi Applications

Numerous examples for real-time Delphi applications can be found. Among them, the Millenium Project conducted by Glenn, Gordon, & Florescu in 2009 provides a context for global thinking and improved understanding of global issues, opportunities, challenges, and strategies. More information on the project can be found on www.stateofthefuture.org

Another stream of projects based on real-time Delphi studies was conducted by the Institute for Futures Studies and Knowledge Management of the EBS Business School in Germany:

1) Future of Logistics – Global Scenarios 2025

2) Transportation & Logistics 2030 – How will supply chains evolve in an energy constrained and low-carbon world

3) The Future of Aviation 2025 – Global Scenarios for Passenger Aviation, Business Aviation and Air Cargo

4) The Indian Aerospace Industry – An Analysis of the Political, Technological and Economic Conditions

5) Transportation & Logistics 2030 – Transport infrastructure – Engine or hand brakes for global supply chains?

## 31.6   References

[1] Hartman, F. T., & Baldwin, A. (1995). Using Technology to Improve Delphi Method. Journal of Computing in Civil Engineering, 9(4), 244-249.

[2] Gordon, T. J., & Pease, A. (2006). RT Delphi: An Efficient, "Round-less", Almost Real Time Delphi Method. Journal of Technological Forecasting and Social Change,73(4), 321-333.

[3] Friedewald, M., von Oertzen, J., & Cuhls, K. (2007). European Perspectives on the Information Society (EPIS) (Delphi Report Deliverable 2.3.1). European Techno-Economic Policy Support Network (ETEPS Net).

[4] Maurizio Bolognini (2001), *Democrazia elettronica: Metodo Delphi e politiche pubbliche (Electronic Democracy: Delphi Method and Public Policy-Making)* (in Italian), Rome: Carocci Editore, ISBN 88-430-2035-8. The author argues that computer-based Delphi can be especially relevant in the context of e-democracy, not only for the roundless approach, the real-time updating of statistical response, or the large number of panellists, but for the possibility to establish split panels corresponding to different groups (such as policy-makers, experts, and citizens), which the administrator can give different tasks and privileges, depending on the issues and the type of decision-making process.

[5] Turoff, M. (1972). Delphi Conferencing: Computer-Based Conferencing with Anonymity. Technological Forecasting and Social Change, 3, 159-204.

[6] Häder, M. (2002). Delphi-Befragungen. Ein Arbeitsbuch. Wiesbaden: Westdeutscher Verlag.

[7] Turoff, M., & Hiltz, S. (1995). Computer based Delphi processes. In M. Adler & E. Ziglio, E. (Eds.), Gazing into the Oracle: The Delphi Method and its Application to Social Policy and Public Health (pp. 56-88). London: Jessica Kingsley Publishers.

[8] Monguet, J., Ferruzca, M., Gutiérrez, A., Alatriste, Y., Martínez, C., Cordoba, C., Fernández, J., et al. (2010). Vector Consensus: Decision Making for Collaborative Innovation Communities. Communications in Computer and Information Science (Vol. 110, pp. 218–227). Viana do Castelo, Portugal: Springer. doi:10.1007/978-3-642-16419-4_22

[9] Gordon, T. J. (2007). Energy forecasts using a "Roundless" approach to running a Delphi study. Foresight, 9(2), 27-35.

[10] Zipfinger, S. (2007). Computer-Aided Delphi: An Experimental Study of Comparing Round-Based with Real-Time Implementation of the Method. Linz: Trauner Verlag.

T. Gnatzy, J.Warth, H.A. von der Gracht, I.-L. Darkow, Validating an Innovative Real-Time Delphi Approach—A methodological comparison between realtime and conventional Delphi studies, Technological Forecasting and Social Change 78 (2011) 1681–1694.

## 31.7 External links

- Institute for Futures Studies and Knowledge Management

# Chapter 32

# Sensorization

**Sensorization** is a modern technology trend to insert many similar sensors in any device or application. Some scientists believe that sensorization is one of main requirements for third technological revolution.[1]

As a result of significant prices drop in recent years there is a trent to include large number of sensors with the same or different function in one device.[2] An example is the evolution of iPhone.

## 32.1   References

[1] Gerard Meijer (2008). *Smart Sensor Systems*. John Wiley & Sons. pp. 1–404. ISBN 978-0-470-86692-4.

[2] "CEA Chief Economist on 2014 Tech Trends Every Retailer Should Watch". *http://www.google.com/think/*. *January 2014. Retrieved 22 January 2014.*

# Chapter 33

# Singularitarianism

**Singularitarianism** is a movement[1] defined by the belief that a technological singularity—the creation of superintelligence—will likely happen in the medium future, and that deliberate action ought to be taken to ensure that the Singularity benefits humans.

Singularitarians are distinguished from other futurists who speculate on a technological singularity by their belief that the Singularity is not only possible, but desirable if guided prudently. Accordingly, they might sometimes dedicate their lives to acting in ways they believe will contribute to its rapid yet safe realization.[2]

*Time Magazine* describes the worldview of Singularitarians by saying that "they think in terms of deep time, they believe in the power of technology to shape history, they have little interest in the conventional wisdom about anything, and they cannot believe you're walking around living your life and watching TV as if the artificial-intelligence revolution were not about to erupt and change absolutely everything." [1]

## 33.1   Alternative definitions

Inventor and futurist Ray Kurzweil, author of the 2005 book *The Singularity Is Near: When Humans Transcend Biology*, defines a Singularitarian as someone "who understands the Singularity and who has reflected on its implications for his or her own life"; he estimates the Singularity will occur around 2045.[2]

## 33.2   History

Singularitarianism coalesced into a coherent ideology in 2000 when artificial intelligence (AI) researcher Eliezer Yudkowsky wrote *The Singularitarian Principles*,[2][3] in which he stated that a "Singularitarian" believes that the singularity is a secular, non-mystical event which is possible and beneficial to the world and is worked towards by its adherents.[3]

In June 2000 Yudkowsky, with the support of Internet entrepreneurs Brian Atkins and Sabine Atkins, founded the Machine Intelligence Research Institute to work towards the creation of self-improving Friendly AI. MIRI's writings argue for the idea that an AI with the ability to improve upon its own design (Seed AI) would rapidly lead to superintelligence. These Singularitarians believe that reaching the Singularity swiftly and safely is the best possible way to minimize net existential risk.

Many people believe a technological singularity is possible without adopting Singularitarianism as a moral philosophy. Although the exact numbers are hard to quantify, Singularitarianism is a small movement, which includes transhumanist philosopher Nick Bostrom. Inventor and futurist Ray Kurzweil, who predicts that the Singularity will occur circa 2045, greatly contributed to popularizing Singularitarianism with his 2005 book *The Singularity Is Near: When Humans Transcend Biology* .[2]

What, then, is the Singularity? It's a future period during which the pace of technological change will be so rapid, its impact so deep, that human life will be irreversibly transformed. Although neither utopian or dystopian, this epoch will transform the concepts we rely on to give meaning to our lives, from our business models to the cycle of human life, including death itself. Understanding the Singularity will alter our perspective on the significance of our past and the ramifications for our future. To truly understand it inherently changes one's view of life in general and one's particular life. I regard someone who understands the Singularity and who has reflected on its implications for his or her own life as a "singularitarian."[2]

With the support of NASA, Google and a broad range of technology forecasters and technocapitalists, the Singularity University opened in June 2009 at the NASA Research Park in Silicon Valley with the goal of preparing the next generation of leaders to address the challenges of accelerating change.

In July 2009, many prominent Singularitarians participated in a conference organized by the Association for the Advancement of Artificial Intelligence (AAAI) to discuss the potential impact of robots and computers and the impact of the hypothetical possibility that they could become self-sufficient and able to make their own decisions. They discussed the possibility and the extent to which computers and robots might be able to acquire any level of autonomy, and to what degree they could use such abilities to possibly pose any threat or hazard (i.e., cybernetic revolt). They noted that some machines have acquired various forms of semi-autonomy, including being able to find power sources on their own and being able to independently choose targets to attack with weapons. They warned that some computer viruses can evade elimination and have achieved "cockroach intelligence." They asserted that self-awareness as depicted in science fiction is probably unlikely, but that there were other potential hazards and pitfalls.[4] Some experts and academics have questioned the use of robots for military combat, especially when such robots are given some degree of autonomous functions.[5] The President of the AAAI has commissioned a study to look at this issue.[6]

## 33.3   Reception

Some critics, such as science journalist John Horgan, compare singularitarianism to religion:

> Let's face it. The singularity is a religious rather than a scientific vision. The science-fiction writer Ken MacLeod has dubbed it "the rapture for nerds," an allusion to the end-time, when Jesus whisks the faithful to heaven and leaves us sinners behind. Such yearning for transcendence, whether spiritual or technological, is all too understandable. Both as individuals and as a species, we face deadly serious problems, including terrorism, nuclear proliferation, overpopulation, poverty, famine, environmental degradation, climate change, resource depletion, and AIDS. Engineers and scientists should be helping us face the world's problems and find solutions to them, rather than indulging in escapist, pseudoscientific fantasies like the singularity.[7]

Kurzweil rejects this categorization, stating that his predictions about the singularity are driven by the data that increases in computational technology have been exponential in the past.[8]

## 33.4   See also

- Friendly AI

- Post scarcity

- Recursive self-improvement

- Strong AI

- Technological singularity

- Outline of transhumanism

## 33.5 References

[1] "2045: The Year Man Becomes Immortal" Time Magazine, February 2011

[2] Kurzweil, Raymond (2005). *The Singularity Is Near: When Humans Transcend Biology*. Viking Adult. ISBN 0-670-03384-7. OCLC 224517172.

[3] Singularitarian Principles"

[4] Scientists Worry Machines May Outsmart Man By John Markoff, NY Times, July 26, 2009.

[5] Call for debate on killer robots, By Jason Palmer, Science and technology reporter, BBC News, 8/3/09.

[6] AAAI Presidential Panel on Long-Term AI Futures 2008-2009 Study, Association for the Advancement of Artificial Intelligence, Accessed 7/26/09.

[7] Horgan, John (2008). "The Consciousness Conundrum". Retrieved 2008-12-17.

[8] Will Google's Ray Kurzweil Live Forever?, Wall Street Journal interview, April 12, 2013

## 33.6 External links

- Ethical Issues in Advanced Artificial Intelligence by Nick Bostrom, 2003

- "The Consciousness Conundrum", a criticism of singularitarians by John Horgan

# Chapter 34

# Technology assessment

**Technology assessment** (**TA**, German *Technikfolgenabschätzung*, French *évaluation des choix scientifiques et technologiques*) is a scientific, interactive, and communicative process that aims to contribute to the formation of public and political opinion on societal aspects of science and technology.[1]

## 34.1  General description

TA is the study and evaluation of new technologies. It is based on the conviction that new developments within, and discoveries by, the scientific community are relevant for the world at large rather than just for the scientific experts themselves, and that technological progress can never be free of ethical implications. Also, technology assessment recognizes the fact that scientists normally are not trained ethicists themselves and accordingly ought to be very careful when passing ethical judgement on their own, or their colleagues, new findings, projects, or work in progress.

Technology assessment assumes a global perspective and is future-oriented, not anti-technological. TA considers its task as interdisciplinary approach to solving already existing problems and preventing potential damage caused by the uncritical application and the commercialization of new technologies.

Therefore, any results of technology assessment studies must be published, and particular consideration must be given to communication with political decision-makers.

An important problem concerning technology assessment is the so-called Collingridge dilemma: on the one hand, impacts of new technologies cannot be easily predicted until the technology is extensively developed and widely used; on the other hand, control or change of a technology is difficult as soon as it is widely used.

Technology assessments, which are a form of cost–benefit analysis, are difficult if not impossible to carry out in an objective manner since subjective decisions and value judgments have to be made regarding a number of complex issues such as (a) the boundaries of the analysis (i.e., what costs are internalized and externalized), (b) the selection of appropriate indicators of potential positive and negative consequences of the new technology, (c) the monetization of non-market values, and (d) a wide range of ethical perspectives.[2] Consequently, most technology assessments are neither objective nor value-neutral exercises but instead are greatly influenced and biased by the values of the most powerful stakeholders, which are in many cases the developers and proponents (i.e., corporations and governments) of new technologies under consideration. In the most extreme view, as expressed by Ian Barbour in "Technology, Environment, and Human Values", technology assessment is "a one-sided apology for contemporary technology by people with a stake in its continuation."[3]

Some of the major fields of TA are: information technology, hydrogen technologies, nuclear technology, molecular nanotechnology, pharmacology, organ transplants, gene technology, artificial intelligence, the Internet and many more. Health technology assessment is related, but profoundly different, despite the similarity in the name.

### 34.1.1 Forms and concepts of technology assessment

The following types of concepts of TA are those that are most visible and practiced. There are, however, a number of further TA forms that are only proposed as concepts in the literature or are the label used by a particular TA institution.[4]

- **Parliamentary TA (PTA)**: TA activities of various kinds whose addressee is a parliament. PTA may be performed directly by members of those parliaments (e.g. in France and Finland) or on their behalf of related TA institutions (such as in the UK, in Germany and Denmark) or by organisations not directly linked to a Parliament (such as in the Netherlands and Switzerland).[5]

- **Expert TA** (often also referred to as the **classical TA** or **traditional TA** concept): TA activities carried out by (a team of) TA and technical experts. Input from stakeholders and other actors is included only via written statements, documents and interviews, but not as in participatory TA.

- **Participatory TA (pTA)**: TA activities which actively, systematically and methodologically involve various kinds of social actors as assessors and discussants, such as different kinds of civil society organisations, representatives of the state systems, but characteristically also individual stakeholders and citizens (lay persons), technical scientists and technical experts. Standard pTA methods include consensus conferences, focus groups, scenario workshops etc.[6] Sometimes pTA is further divided into **expert-stakeholder pTA** and **public pTA** (including lay persons).[7]

- **Constructive TA (CTA)**: This concept of TA, developed in the Netherlands, but also applied and discussed elsewhere[8] attempts to broaden the design of new technology through feedback of TA activities into the actual construction of technology. Contrary to other forms of TA, CTA is not directed toward influencing regulatory practices by assessing the impacts of technology. Instead, CTA wants to address social issues around technology by influencing design practices.

- **Discursive TA** or **Argumentative TA**: This type of TA wants to deepen the political and normative debate about science, technology and society. It is inspired by ethics, policy discourse analysis and the sociology of expectations in science and technology. This mode of TA aims to clarify and bring under public and political scrutiny the normative assumptions and visions that drive the actors who are socially shaping science and technology. Accordingly, argumentative TA not only addresses the side effects of technological change, but deals with both broader impacts of science and technology and the fundamental normative question of why developing a certain technology is legitimate and desirable.[9]

- **Health TA (HTA)**: A specialised type of expert TA informing policy makers about efficacy, safety and cost effectiveness issues of pharmaceuticals and medical treatments, see health technology assessment.

## 34.2 Technology assessment institutions around the world

Many TA institutions are members of the European Parliamentary Technology Assessment (EPTA) network, some are working for the STOA panel of the European Parliament and formed the European Technology Assessment Group (ETAG).

- Centre for Technology Assessment (TA-SWISS), Bern, Switzerland.

- Institute of Technology Assessment (ITA) of the Austrian Academy of Sciences, Vienna

- Institute for Technology Assessment and Systems Analysis, Karlsruhe Institute of Technology, Germany

- (former) Office of Technology Assessment (OTA)

- The Danish Board of Technology Foundation, Copenhagen

- Norwegian Board of Technology, Oslo

- Parliamentary Office of Science and Technology (POST), London

- Rathenau Institute, The Hague

- Science and Technology Options Assessment (STOA) panel of the European Parliament, Brussels

- Science and Technology Policy Research (SPRU), Sussex

- t.b.c.

## 34.3   See also

- Collingridge dilemma

- Technology

- Technology dynamics

- Technology forecasting

- Technology readiness level

- Technology transfer

## 34.4   External links

- Scientific Technology Options Assessment (STOA), European Parliament

- European Technology Assessment Group for STOA

- Institute for Technology Assessment and Systems Analysis (ITAS), Karlsruhe Institute of Technology (KIT), Germany

- Office of Technology Assessment at the German Parliament (TAB)

- TA-SWISS Centre for Technology Assessment

- Institute of Technology Assessment (ITA), Austrian Academy of Sciences, Vienna, Austria

- The Danish Board of Technology

- Rathenau Institute

- The Norwegian Board of Technology

## 34.5   References

[1] Cf. the commonly used definition given in the report of the EU-funded project TAMI (Technology Assessment – Methods and Impacts) in 2004: ta-swiss.ch

[2] Huesemann, Michael H., and Joyce A. Huesemann (2011). *Technofix: Why Technology Won't Save Us or the Environment*, Chapter 8, "The Positive Biases of Technology Assessments and Cost Benefit Analyses", New Society Publishers, Gabriola Island, British Columbia, Canada, ISBN 0865717044, 464 pp.

[3] Barbour, I.A. (1980). *Technology, environment, and human values*, Praeger, p. 202.

[4] Among those concepts one finds, for instance, **Interactive TA** ITAS.fzk.de, **Rational TA** EA-AW.com, **Real-time TA** (cp. Guston/Sarewitz (2002) Real-time technology assessment, in: Technology in Society 24, 93–109), Innovation-oriented TA Innovationsanalysen.

[5] Those TA institutions that perform PTA are organised in the European Parliamentary Technology Assessment (EPTA) network; see EPTAnetwork.org.

[6] Cp. the 2000 EUROpTA (European Participatory Technology Assessment – Participatory Methods in Technology Assessment and Technology Decision-Making) project report TEKNO.dk.

[7] Van Eijndhoven (1997) Technology assessment: Product or process? in: Technological Forecasting and Social Change 54 (1997) 269–286.

[8] Schot/Rip (1997), The Past and Future of Constructive Technology Assessment in: Technological Forecasting & Social Change 54, 251–268.

[9] van Est/Brom (2010) Technology assessment as an analytic and democratic practice, in: Encyclopedia of Applied Ethics.

## 34.6 Text and image sources, contributors, and licenses

### 34.6.1 Text

- **Technology forecasting** *Source:* https://en.wikipedia.org/wiki/Technology_forecasting?oldid=673620494 *Contributors:* Anthere, Ronz, Altenmann, Vadmium, Jareha, Wikiacc, Kross, Bobo192, Enric Naval, Pearle, Woohookitty, RHaworth, Hulagutten, Yamamoto Ichiro, Nihiltres, AndriuZ, Arthur Rubin, SmackBot, Reedy, Hmusseau, Silly rabbit, Colonies Chris, Jorvik, Tekno4caster, Kuru, Hu12, Gregbard, Dancter, Spearman, The Transhumanist, Maqayum, Longly, Avicennasis, Apdevries, R'n'B, Rjhyndman, Flyer22 Reborn, Sphilbrick, Rohrbeck, SlackerMom, Bob bobato, Trivialist, 7, Cheapskate08, Addbot, MrOllie, Glane23, Yobot, Chyen, Citation bot, Abcdeeeee8, RibotBOT, Eliovir, Thehelpfulbot, Mnent, Tom.Reding, Bill Halal, Minimac, Lizzle0909, John of Reading, AsceticRose, DrScholar, 52andrew, Econ2010, Tideflat, Jeraphine Gryphon, Marcoantonioacostareyes, Mbettersworth, Lemnaminor, Monkbot and Anonymous: 34

- **2081: A Hopeful View of the Human Future** *Source:* https://en.wikipedia.org/wiki/2081%3A_A_Hopeful_View_of_the_Human_Future?oldid=662145852 *Contributors:* Alan Liefting, Wronkiew, A.T.M.Schipperijn, GregorB, Fresheneesz, Wavelength, Pegship, SmackBot, Chris the speller, Cattus, John, The Enslaver, Danrok, HolyT, VoABot II, Useight, Macy, Howie Goodell, Nonegivenwp, LivingBot, CobraBot, Werieth, Midas02, Helpful Pixie Bot, Skyhook1 and Anonymous: 6

- **5G** *Source:* https://en.wikipedia.org/wiki/5G?oldid=689248917 *Contributors:* Michael Hardy, DmitryKo, Discospinster, Andros 1337, StephanKetz, Jlin, Etrigan, Wtmitchell, DePiep, Rjwilmsi, Trlovejoy, Naraht, Bhny, Cwlq, Stephenb, Bachrach44, Arthur Rubin, Mhenriday, SmackBot, Alistair9210, Jenny MacKinnon, Frap, KevM, Khazar, Oioisaveloy, Kashmiri, Melody Concerto, Mark999, Kvng, Hollomis, Ice Ardor, MERC, XAKxRUSx, Hekerui, David Eppstein, An Sealgair, Jim.henderson, R'n'B, Mange01, Acalamari, Shadowaltar, Psheld, Jpgs, GcSwRhlc, Lordvolton, WereSpielChequers, Flyer22 Reborn, Jimthing, Sfan00 IMG, Plastikspork, Supremedemency, Muhandes, Arjayay, InternetMeme, Amitnaik, XLinkBot, MystBot, Bazj, Addbot, C933103, Luckas-bot, Yobot, Ptbotgourou, KamikazeBot, Langthorne, AnomieBOT, Ulric1313, Materialscientist, DynamoDegsy, Xqbot, Anna Frodesiak, SilverSurfer477, Shadowjams, A.amitkumar, Ch Th Jo, Nageh, Sanpitch, Jonesey95, Robvanvee, Trappist the monk, Mean as custard, RjwilmsiBot, EmausBot, John of Reading, WikitanvirBot, Dewritech, RA0808, ZéroBot, Illegitimate Barrister, Thayora, Ftprinc, ChuispastonBot, Marcelocantos, Cgt, ClueBot NG, Widr, Helpful Pixie Bot, Novusuna, Imthiyaz5g, BG19bot, AvocatoBot, Civeel, Suryaips, Andreas.kagedal, Piano1900, Demsiti, Hamish59, Sni56996, Klilidiplomus, Theemstra, BattyBot, 3enix, Cleliasayswhat, John from Idegon, Jules Brunet, EuroCarGT, Alxsnc12, TomoK12, Mogism, Digitalcrowd, Frosty, MrCellular, Vanamonde93, Psatsankhya, Jakec, DavidLeighEllis, Comp.arch, Wiskica, Melody Lavender, D Eaketts, Jomondel, SJ Defender, Dario.sabella, Wow560, Tamersaadeh, Drvipinlalt, Wyn.junior, 22merlin, Jackiecsq123, DrBoris1, Ebonelm, Cheval55, Alex1961, Manunamboodiri, Shahzad.tut, Aytk, Jtxxtj, Alimxd1, Alexm38, Enkakad, Infopaktel, Sunilkumargc, Jerodlycett, Sridhar vasanth sri, Hafiz Badar Zubair, Hajisamik, Crazy crayon, Jbmorman and Anonymous: 196

- **Big data** *Source:* https://en.wikipedia.org/wiki/Big_data?oldid=689300774 *Contributors:* William Avery, Heron, Kku, Samw, Andrewman327, Tpbradbury, Ryuch, דוד, Topbanana, Paul W, F3meyer, Sunray, Giftlite, Langec, Erik Carson, Utcursch, Beland, Jeremykemp, David@scatter.com, Discospinster, Rich Farmbrough, Kdammers, ArnoldReinhold, Narsil, Viriditas, Lenov, Gary, Pinar, Tobych, Miranche, Broeni, Tomlzz1, Axeman89, Woohookitty, Pol098, Qwertyus, Rjwilmsi, Koavf, ElKevbo, Jehochman, Nihiltres, Lumin~enwiki, Tedder, DVdm, SteveLoughran, Aeusoes1, Daniel Mietchen, Cedar101, Dimensionsix, Katieh5584, Henryyan, McGeddon, Od Mishehu, Gilliam, Ohnoitsjamie, Chris the speller, RDBrown, Pegua, Madman2001, Krexer, Kuru, Almaz~enwiki, DI2000, The Letter J, Chris55, Yragha, Jac16888, Marc W. Abel, Cydebot, Matrix61312, Quibik, DumbBOT, Malleus Fatuorum, EdJohnston, Nick Number, Cowb0y, Lmusher, Josephmarty, Kforeman1, Rmyeid, OhanaUnited, Relyk, Wllm, Lvsubram, Magioladitis, Nyq, Tedickey, Steven Walling, Thevoid00, Casieg, Jim.henderson, Tokyogirl79, MacShimi, McSly, NewEnglandYankee, Lamp90, Asefati, Pchackal, Mgualtieri, VolkovBot, JohnBlackburne, Vincent Lextrait, Philip Trueman, Ottb19, Billinghurst, ParallelWolverine, Grinq, Scottywong, Luca Naso, Dawn Bard, Yintan, Jazzwang, Jojikiba, Eikoku, SPACKlick, CutOffTies, Mkbergman, Melcombe, Siskus, PabloStraub, Dilaila, Martarius, Sfan00 IMG, Faalagorn, Apptrain, Morrisjd1, Grantbow, Mild Bill Hiccup, Ottawahitech, Cirt, Auntof6, Lbertolotti, Gnome de plume, Resoru, Pablomendes, Saisdur, Vehementlyirish, SchreiberBike, MPH007, Rui Gabriel Correia, Mymallandnews, XLinkBot, Ost316, Benboy00, MystBot, Itadapter, P.r.newman, Addbot, Mortense, Drevicko, Thomas888b, Non-dropframe, AndrewHZ, Tothwolf, Ronhjones, Moosehadley, MrOllie, Download, Jarble, Arbitrarily0, Luckas-bot, Yobot, Fraggle81, Manivannan pk, Misterlevel, Elfix, Jean.julius, AnomieBOT, Jim1138, Babrodtk, Bluerasberry, Materialscientist, Citation bot, Xqbot, Marko Grobelnik, Melmann, Bgold12, Anna Frodesiak, Tomwsulcer, Srich32977, Omnipaedista, Smallman12q, Joaquin008, Jugdev, FrescoBot, Jonathanchaitow, I42, PeterEastern, AtmosNews, B3t, I dream of horses, HRoestBot, Jonesey95, Jandalhandler, Mengxr, Ethansdad, Yzerman123, Msalganik, בן גרשון, Sideways713, Stuartzs, Jfmantis, Mean as custard, RjwilmsiBot, Ripchip Bot, Mm479flarok, Winchetan, Petermcelwee, DASHBot, EmausBot, John of Reading, Oliverlyc, Timtempleton, Dewritech, Peaceray, Radshashi, Cmlloyd1969, K6ka, HiW-Bot, Richard asr, ZéroBot, Checkingfax, BobGourley, Josve05a, Xtzou, Chire, Kilopi, Laurawilber, Rcsprinter123, Rick jens, Palosirkka, MainFrame, ChuispastonBot, Sean Quixote, Axelode, Mhiji, Helpsome, ClueBot NG, Behrad3d, Danielg922, Pramanicks, Jj1236, Widr, WikiMSL, Lawsonstu, Fvillanustre, Helpful Pixie Bot, Lowercase sigmabot, BG19bot, And Adoil Descended, Seppemans123, Jantana, Innocentantic, Northamerica1000, Asplanchna, MusikAnimal, AvocatoBot, Noelwclarke, Matt tubb, Jordanzhang, Bar David, InfoCmplx, Atlasowa, Fylbecatulous, Camberleybates, BattyBot, WH98, DigitalDev, Haroldpolo, Ryguyrg, Untioencolonia, Shirishnetke, Ampersandian, MarkTraceur, ChrisGualtieri, TheJJJunk, Khazar2, Vaibhav017, IjonTichyIjonTichy, Saturdayswiki, Mheikkurinen, Seherrell, Mjvaugh2, Chazz173, Davidogm, Dexbot, Mherradora, Jkofron4, Stevebillings, Indianbusiness, Toopathfind, Jeremy Kolb, Frosty, Jamesx12345, OnTheNet21, BrighterTomorrow, Phamnhatkhanh, Jacoblarsen net, Epicgenius, DavidKSchneider, Socratesplato9, Anirudhrata, Parasdoshiblog, Edwinboothnyc, JuanCarlosBrandt, Helenellis, MMeTrew, Warrenpd86, Michael.alexander.kaufmann, AuthorAnil, ViaJFK, Gary Simon, Bsc, FCA, FBCS, CITP, Mcioffi, Joe204, Caraconan, Evaluatorgroup, Hessmike, TJLaher123, Chenying10, IndustrialAutomationGuru, Dabramsdt, Prussonyc, Abhishek1605, Dilaila123, Willymomo, Rzicari, Mandruss, Mingminchi, BigDataGuru1, Sugamsha, Sysœp, Azra2013, Paul2520, Dudewhereismybike, Shahbazali101, SJ Defender, Yeda123, Miakeay, Stamptrader, Accountdp, Morganmissen, JeanneHolm, Yourconnotation, JenniferAndy, Arcamacho, Amgauna, Bigdatavomit, Monkbot, Wikientg, Scottishweather, Textractor, Analytics ireland, Lspin011, ForumOxford Online, Mansoor-siamak, Belasobral, Sightestrp, Jwdang4, Amortias, Wikiauthor22, Femiolajiga, Tttcraig, Lepro2, Mythfinder, DexterToo, Mr P. Kopee, Pablollopis, SVtechie, Deathmuncher19, Smaske, Greystoke1337, Prateekkeshari, Hmrv83, Vidyasnap, KaraHayes, Iqmc, Lalith269, Helloyoubum, Jakesher, IEditEncyclopedia, Rajsbhatta123, Ragnar Valgeirsson, Vedanga Kumar, Fgtyg78, Gary2015, HelpUsStopSpam, EricVSiegel, Benedge46, Friafternoon, KasparBot, Adzzyman, Pmaiden, Spetrowski88, JuiAmale, Yasirsid, Diyottainc, Nt8068a, WikilleWi, Preyansh07, Dharnett21 and Anonymous: 364

- **Big Memory** *Source:* https://en.wikipedia.org/wiki/Big_Memory?oldid=687420196 *Contributors:* Itadapter, Leo gan 57, BG19bot, Rberchie, Jj14, HelpUsStopSpam and Anonymous: 1

- **Center for Nanotechnology in Society** *Source:* https://en.wikipedia.org/wiki/Center_for_Nanotechnology_in_Society?oldid=682665894 *Contributors:* Ricky81682, Rjwilmsi, Ethansmith, The Anomebot2, Cshearer19, Citation bot, Jonesey95, Tbhotch, Josve05a and Anonymous: 1

- **Center on Nanotechnology and Society** *Source:* https://en.wikipedia.org/wiki/Center_on_Nanotechnology_and_Society?oldid=552896619 *Contributors:* TiMike, RoyBoy, Quiddity, Alphachimp, SpuriousQ, 建建建 robot, SmackBot, Jonrhodes18, MarshBot, Erechtheus, The Anomebot2, DGG, D Invictus, Tikiwont, Ypetrachenko, Kyriosity, Qst, Joe8824, John of Reading, Insomesia and Anonymous: 4

- **Clarke's three laws** *Source:* https://en.wikipedia.org/wiki/Clarke'{}s_three_laws?oldid=688769720 *Contributors:* Damian Yerrick, Axel-Boldt, Bryan Derksen, The Anome, Pinkunicorn, DavidLevinson, Modemac, Tillwe, Michael Hardy, Dougmerritt, Paul A, Kosebamse, Tregoweth, Mxn, Arteitle, PaulinSaudi, Furrykef, Fibonacci, Sdedeo, Xanzzibar, Mattflaschen, DavidCary, Agentseven, Guanaco, Siroxo, Tagishsimon, Mdob, Antandrus, Piotrus, Khaosworks, Malakhi, Pat Berry, Kuralyov, Neutrality, Julianonions, Ben Zealley, JasticE, Jimaginator, Grstain, Rich Farmbrough, Guanabot, Notinasnaid, Kaisershatner, Elwikipedista~enwiki, Kwamikagami, Mqduck, Jpgordon, AmosWolfe, Spoom, Rajah, DariuszT, Andrew Gray, Linmhall, Ferrierd, Juri, RJFJR, Dirac1933, Mindmatrix, Oliphaunt, Tabletop, Crazysunshine, Radiant!, Keeves, Tlroche, JWWalker, The wub, Yar Kramer, Jacobus71, Thegreatloofa, Nickpowerz, Alphachimp, DerrickOswald, Wasted Time R, Michael Slone, Bachrach44, Caiyu, CLW, Crumley, Pawyilee, Thnidu, Streltzer, JonTown, Garion96, Sneftel, Kingboyk, Attilios, SmackBot, Eperotao, FlashSheridan, Vald, GraemeMcRae, Squiddy, Manuelomar2001, Thumperward, ERobson, Petermr, Huon, Dacoutts, Bostwickenator, MSchmahl, BillFlis, George The Dragon, Tihnessa, FatBastardInk, CmdrObot, Artem82, N2e, Racooper, Gregbard, Legendary, Dogman15, Dennette, Goldfritha, Tenbergen, MMoudry, Raistlin Majere, Rocket000, Roguebfl, Thijs!bot, Fbe2, MainlyTwelve, Noclevername, Spectheintro, QuiteUnusual, Albany NY, Roccondil, .Absolution., VoABot II, Antipodean Contributor, Svunt, Barry Haworth, Gwern, Misarxist, CFCF, OttoMäkelä, Lantonov, Carolfrog, Chaveyd, Richard D. LeCour, Dhaluza, STBotD, Master z0b, N3ug3, Tuba mirum, JcsherwoodSH1854, Poketape, LrdSlvrhnd, SheffieldSteel, MWielage, DanonCanon, Alhead, Paradoctor, GeiwTeol, Nervousenergy, Doctorfluffy, Svick, Skeptical scientist, Gorkymalorki, Randy Kryn, Martarius, Tal9922, Homonihilis, Three-quarter-ten, Resoru, PixelBot, PaulxSA, GreenGorgon, XLinkBot, Kolyma, Kolacinski, Couzin2000~enwiki, Stephen Poppitt, Gsagers, Download, SpBot, Legobot, Luckas-bot, MileyDavidA, Yobot, AnomieBOT, DoctorJoeE, LilHelpa, Koyarpm, Hydrox24, Flying sheep, NobelBot, Lost tiree, lost dutch :O, FrescoBot, Jonesey95, Skyerise, Jonathan.h.brown, Miracle Pen, Gadflyr, Metaferon, EmausBot, AmigoCgn, Schwa dk, Slightsmile, ZéroBot, Ethaniel, Ό οἶστρος, Brc2000, Western John, ClueBot NG, Despatche, Dingowasher, Reify-tech, Farragoer, PseudoImmortal, BG19bot, AnEyeSpy, Tycho Magnetic Anomaly-1, Cbargren, ChrisGualtieri, Mr. Guye, Fizvo, Tango303, Opencooper, Livvysoda147 and Anonymous: 210

- **Clean Energy Trends** *Source:* https://en.wikipedia.org/wiki/Clean_Energy_Trends?oldid=627722444 *Contributors:* Alan Liefting, Wavelength, Ckatz, Harryzilber, Chris G, Johnfos, MrOllie, Htomfields, Dpsypher, Appeltree1, Mdavidson98 and Anonymous: 4

- **Datafication** *Source:* https://en.wikipedia.org/wiki/Datafication?oldid=638433518 *Contributors:* Deb, Alexdruk and Tristan Sterk

- **Delphi method** *Source:* https://en.wikipedia.org/wiki/Delphi_method?oldid=680240828 *Contributors:* Michael Hardy, Ronz, Nikai, Charles Matthews, Reddi, Zoicon5, Robbot, Paranoid, Thunderbolt16, Craigwb, Angry candy, MichaelHaeckel, Jeremykemp, Binand, Smokris, Madduck, Mani1, Night Gyr, Kwamikagami, Bobo192, Flammifer, Mdd, Malo, Shoefly, Crosbiesmith, Qwertyus, Rjwilmsi, Jivecat, Wiarthurhu, Dmccreary, Tequendamia, YurikBot, Argav, Stephenb, Gaius Cornelius, Chriswaterguy, JLaTondre, FritzSolms, SmackBot, Jtneill, Mauls, Gilliam, Kukini, SashatoBot, Meco, Cydebot, Naudefj, Cs california, Thijs!bot, Wikid77, Andyjsmith, Bobblehead, Ekalaivan, Deflective, Skomorokh, Ph.eyes, Dscotese, Shedai, DRHagen, James S. Friedman, Omicron18, WhatamIdoing, Seleucus, Apdevries, R'n'B, Commons-Delinker, RJBurkhart3, Pilgaard, Kimleonard, Kesten~enwiki, Warut, VolkovBot, Bbik, Rescherpa, Briansuda, To what end, Metasailor, AlleborgoBot, SieBot, Melcombe, ClueBot, ImperfectlyInformed, Agraefe, DaveBurstein, Excirial, Mhhfive, Vegetator, Qwfp, Editor2020, Tayste, Addbot, Norman21, Lightbot, Legobot, Yobot, The onlooker, THEN WHO WAS PHONE?, Nallimbot, AnomieBOT, DemocraticLuntz, JackieBot, Kingpin13, Xqbot, GrouchoBot, FrescoBot, GEBStgo, TobeBot, Dinamik-bot, Scolton, Tavennerfb, Dps04, Makki98, Lizzle0909, EmausBot, Solarra, Petapio, Philphilphilphil, Esaintpierre, Pokbot, Will Beback Auto, ClueBot NG, Czkssa, Helpful Pixie Bot, Elby Team, Hans Frörum, ProjectManhattan, Ndr76, CitationCleanerBot, Sailor7sakura, Sa publishers, Oracle time, Monkbot, Fletambe, JohnElliotV, Walwany and Anonymous: 136

- **Differential technological development** *Source:* https://en.wikipedia.org/wiki/Differential_technological_development?oldid=652113855 *Contributors:* David Gerard, Centrx, Loremaster, Bookofjude, GregorB, RussBot, Banus, Phuzion, Addbot, AnomieBOT, Crzer07, Tom.Reding, EmausBot, Brian Tomasik, BattyBot, Monkbot and Anonymous: 5

- **Emerging technologies** *Source:* https://en.wikipedia.org/wiki/Emerging_technologies?oldid=686700853 *Contributors:* Edward, Haakon, Reddi, Greenrd, Sunray, David Gerard, Andycjp, Alexf, Loremaster, Discospinster, FT2, Vsmith, Pgptag, Slambo, Gary, Mikeo, SteinbDJ, Oleg Alexandrov, Bobrayner, Firsfron, Mandarax, Hulagutten, Eyu100, ElKevbo, SchuminWeb, Ksyrie, Modal Auxiliary, Wangi, Closedmouth, Kingkiki217, SmackBot, Tigerghost, McGeddon, Gilliam, Chris the speller, Andrew c, Metamagician3000, Kuru, Nakedtruth, MaxEnt, Cydebot, Crossmr, Thomas Paine1776, Wayiran, Barek, The Transhumanist, VoABot II, Zedomax, Pax:Vobiscum, Rickard Vogelberg, R'n'B, PStrait, J.delanoy, Mange01, Karanacs, Kar.ma, Icseaturtles, Mikael Häggström, Fountains of Bryn Mawr, Lifeboatpres, Bureb62, Tcgriffin, Wessmaniac, Agalmic, Martarius, ClueBot, Bradka, Niceguyedc, Excirial, Yorkshirian, 1ForTheMoney, BendersGame, XLinkBot, Jovianeye, Avoided, Kbdankbot, Addbot, Quercus solaris, LarryJeff, Yobot, Thovonnc7, AnomieBOT, Samuel Lann, Materialscientist, Citation bot, CoolingGibbon, DSisyphBot, Gap9551, GrouchoBot, Solphusion~enwiki, Alfredmiller, MerlLinkBot, A.amitkumar, Thierrynet, Paine Ellsworth, Jonatech, Pinethicket, Edderso, 10metreh, Tom.Reding, Chabi1, Metatr0n108, Tucvbif, Juhko, Mono, Mean as custard, Lizzle0909, Chakamull, Wayne Slam, Classic mechanic, MonoAV, Damirgraffiti, Rangoon11, ClueBot NG, MelbourneStar, Tideflat, Rasebastian, Vivaystn, Helpful Pixie Bot, Jeraphine Gryphon, BG19bot, Thigalepranav, Claviere, Iyj 1991, Kodip, Graphenebiz, Melonkelon, Mcioffi, EvergreenFir, Rambo FT, Bordwall, EJET63, Dportzline, Neatsfoot, KristinaMcloud, Amariiona, Rise of Technology, Junaid sipra, Danielerotolo, Infinite0694, Prajotpatil, Wellspc, Glb613 and Anonymous: 109

- **The Future of the Mind** *Source:* https://en.wikipedia.org/wiki/The_Future_of_the_Mind?oldid=635546327 *Contributors:* Bearcat, Kateshortforbob, Lamro, TheLongTone, Linuxjava, Twhitguy14ma and Anonymous: 1

- **Futures studies** *Source:* https://en.wikipedia.org/wiki/Futures_studies?oldid=688114956 *Contributors:* Bryan Derksen, The Anome, Michael Hardy, Ronz, Reddi, Jerzy, Robbot, Tomchiukc, Gandalf61, Pmcray, David Gerard, Geeoharee, Bfinn, Curps, Alison, Pasquale, Brianhe, Florian Blaschke, Bender235, Jnestorius, Guettarda, Nectarflowed, Viriditas, L33tminion, Giraffedata, Jeodesic, Mdd, BDD, Drbreznjev, Mark K. Jensen, BD2412, Josh Parris, Drbogdan, Chobot, DVdm, Yamara, Manop, Boivie, Anclation~enwiki, TLSuda, Snalwibma, SmackBot, Vald, Facius, Stephensuleeman, Ppntori, Rielm, Nedlum, Scwlong, Rhodesh, RJN, "alyosha", Sethwoodworth, TenPoundHammer, Will Beback, Ckatz, Caiaffa, JohnSmart, Woodroar, Twas Now, Shammaee, CRGreathouse, Nova325, Neelix, Ksoileau, Gregbard, Yaris678, Cydebot, Treybien, Futureobservatory, Dancter, Doug Weller, Maziotis, Epbr123, Lynndunn, Dfrg.msc, Widefox, Blue Tie, Fayenatic london, Pikle, JAnDbot, The Transhumanist, Michig, John b cassel, Boleslaw, .anacondabot, Lenny Kaufman, Mclay1, Japo, Falcor84, Kayau, DGG, FisherQueen, CommonsDelinker, RJBurkhart3, Shaunfensom, Wikip rhyre, DadaNeem, 83d40m, Toon05, Joshua Issac, FuegoFish, Bonadea, Ferryiti, DASonnenfeld, Squids and Chips, Funandtrvl, VolkovBot, Pelarmian, Rescherpa, Patrizio2, AllGloryToTheHypnotoad, UnitedStatesian, Jamelan, Meters, Spinningspark, Monty845, Sirswindon, Thepaperbicycle, Tresiden, Nihil novi, ToePeu.bot, Dawn Bard, Doctorfluffy, Sphilbrick, Carl weathers bicep, Ijbaker, EmanWilm, Rohrbeck, Martarius, Cyberlisa, DragonBot, Howie Goodell, Kitsunegami, Excirial, Srineev, Martinkrusemrk, Byoboo, Tired time, Thingg, ThisMunkey, Chipmunker, Editor2020, DumZiBoT, XLinkBot, Martinkruse07, Leoniana, Jonathanmoyer, Mabalu, Hawkania, Addbot, Grayfell, Sohail Inayatullah, DOI bot, Neodop, Anders Sandberg, Yelizandpaul, RTG, Debresser, Favonian, OlEnglish, Lumberjack89, Yobot, Ptbotgourou, TaBOT-zerem, Shimk83, Vini 17bot5, Jcanton, AnomieBOT, Rubinbot, ThaddeusB, Jim1138, Chyen, Hunnjazal, Thepogoman, LilHelpa, Obersachsebot, Xqbot, Gotophilk, Abcdeeeee8, Gap9551, AdamGordon, GrouchoBot, Alewishines, Hifcelik, Omnipaedista, Mark Schierbecker, CHJL, Wmcg2, Stephen Grandt, Mnent, CaptainFugu, Anssi tervonen, Mindaisy, Abductive, Szea426, Hellknowz, SpaceFlight89, Wfsf, Intelligentelf, SkyMachine, Idoforgod, NimbusWeb, GregKaye, Amar2556, Indy1981, Mramz88, Tobycumming, BrightBlackHeaven, J Shirk, Alph Bot, EmausBot, John of Reading, GoingBatty, TeleComNasSprVen, Thryller, Bethnim, Susfele, H3llBot, DrScholar, Noodleki, Philphilphilphil, Lampsalot, Smith2200, ClueBot NG, Pandelver, StatPak, Mahir256, Wavefront3, Christopher0671, Responder55, Cybernew, Joel B. Lewis, Sprints, MerlIwBot, Meximore, Helpful Pixie Bot, Jeraphine Gryphon, Soroush90gh, BG19bot, PhnomPencil, Writ Keeper, Mejoribus, Rextroumbley, The Illusive Man, Myxomatosis57, Peterjungk, Khazar2, Karenmariearvidsson, Tyler Hruby, TravellingThru, Delphenich, Ahadrezayan, Dustin V. S., Fluous, LudwidNDes, Prokaryotes, Climate123, Emiruzunoglu, Cendrake, Linuxjava, Fixuture, Cmendoza67, Monkbot, Aarvonen, GinAndChronically, Ally Oxford, Omarsahi, Khenzackenz022, Jalfredpeacock, ChamithN, Linskon, ScrapIronIV, Anadrev, JJS42, Mphelan2015, Compassionate727, Acowart10, Oalikor, Maxaxax and Anonymous: 197

- **Futuribles International** *Source:* https://en.wikipedia.org/wiki/Futuribles_International?oldid=579930949 *Contributors:* FayssalF, Dialectric, True Pagan Warrior, SmackBot, Cydebot, Xhienne, Addbot and ChrisGualtieri

- **Ray Hammond** *Source:* https://en.wikipedia.org/wiki/Ray_Hammond?oldid=633524503 *Contributors:* Mattstan, Pinball22, Mandarax, Bgwhite, Gholam, Courcelles, Cydebot, Deadbeef, Waacstats, Martarius, Adrianwn, Addbot, Push the button, RjwilmsiBot, Comatmebro and Anonymous: 4

- **I = PAT** *Source:* https://en.wikipedia.org/wiki/I_%3D_PAT?oldid=687741883 *Contributors:* Timo Honkasalo, Psi~enwiki, Tannin, Ixfd64, Peregrine981, AaronSw, Seth Ilys, Alan Liefting, Everyking, Onco p53, CALR, Rich Farmbrough, Vsmith, Bender235, Woohookitty, Zrenneh, Commander Keane, BD2412, Volunteer Marek, Bgwhite, Gareth Jones, Arthur Rubin, AlexD, SmackBot, Incnis Mrsi, Scientizzle, Dr.enh, LRG, Mack2, Steveprutz, Gabriel Kielland, DASonnenfeld, Agricola44, ClueBot, Shortspecialbus, Zodon, Jarble, Legobot, Yobot, AnomieBOT, Jim1138, Materialscientist, M zhou7, FrescoBot, Tom.Reding, GregKaye, RjwilmsiBot, Josve05a, Jonpatterns, ClueBot NG, Greenpea vn, Dexbot, Reynardloki, Alpha Sigma 111, MichaelAGallo, Waters.Justin, User000name, Stewi101015 and Anonymous: 58

- **Immersion (virtual reality)** *Source:* https://en.wikipedia.org/wiki/Immersion_(virtual_reality)?oldid=686238488 *Contributors:* Frecklefoot, Haakon, Bevo, Robbot, Alan Liefting, Alexf, Bender235, Grutness, Diego Moya, Woohookitty, BD2412, Rjwilmsi, Wavelength, Dialectric, Welsh, SmackBot, Edgar181, Ohnoitsjamie, Fluss, GVnayR, Kukini, Optakeover, Aeternus, J Milburn, Cydebot, Alaibot, EWAdams, Jodi.a.schneider, CommonsDelinker, SharkD, LordAnubisBOT, Nwbeeson, Funandtrvl, Horatiorama, ImageRemovalBot, Lordhood117, Martarius, Vladkornea, XLinkBot, Addbot, Non-dropframe, Psynapz, MrOllie, Applesweet6, SpBot, Legobot, Luckas-bot, Yobot, Taxisfolder, AnomieBOT, Valueyou, Brightgalrs, LilHelpa, Bringmetheheadofalfredogarcia, DSisyphBot, Measles, Crzer07, Solphusion~enwiki, Лев Дубовой, DrilBot, Lotje, EmausBot, John of Reading, Playmobilonhishorse, Midas02, SporkBot, AManWithNoPlan, Rangoon11, ClueBot NG, Krunchyman, CasualVisitor, Helpful Pixie Bot, CrimLawProf, Virtualerian, Espejo s, MindTrainingOnline, BattyBot, Hansen Sebastian, ChrisGualtieri, Dexbot, Faizan, Thomys~enwiki, Wakey0103, Heraclites, Monkbot, Vieque, Babymyggan, BrettReaper and Anonymous: 54

- **Industrial Big Data** *Source:* https://en.wikipedia.org/wiki/Industrial_Big_Data?oldid=689043395 *Contributors:* Behrad3d, BG19bot and Anonymous: 1

- **Industrial Internet** *Source:* https://en.wikipedia.org/wiki/Industrial_Internet?oldid=687823462 *Contributors:* Ttwaring, Wavelength, Derek R Bullamore, SamShearman, Addbot, Jarble, Yobot, AnomieBOT, A.wasylewski, Behrad3d, Wbm1058, BG19bot, ChrisGualtieri, Bezoma, ViaJFK, Nick Stavros, Sattar91, IndustrialAutomationGuru, Saectar, Mr P. Kopee, Fellmark, Jinco, Esthan and Anonymous: 11

- **Predix (software)** *Source:* https://en.wikipedia.org/wiki/Predix_(software)?oldid=686535988 *Contributors:* Chikyuu, Jcronen1, SwisterTwister, GoingBatty, BG19bot and Sattar91

- **Industry 4.0** *Source:* https://en.wikipedia.org/wiki/Industry_4.0?oldid=684923962 *Contributors:* Ronz, M0mms, Oneiros, Giraffedata, Welsh, Michael.kenward, Fram, Voceditenore, Tmangray, Nick Number, Magioladitis, Bassplr19, Namazu-tron, Tassedethe, Yobot, AnomieBOT, BobKilcoyne, Horst-schlaemma, Hanky27, ClueBot NG, Behrad3d, Croesomorris, Wbm1058, BG19bot, Fex80~enwiki, MeanMotherJr, ChrisGualtieri, Eyesnore, Murmelsson, Lagoset, HgPchrd, Fellmark, Mbsnom, Keith Vincent Lester, Okbanduss and Anonymous: 29

- **Kaya identity** *Source:* https://en.wikipedia.org/wiki/Kaya_identity?oldid=678937943 *Contributors:* Edward, William M. Connolley, SEWilco, Aetheling, Alan Liefting, Thierryc, Bobrayner, Rjwilmsi, DF5GO, Bhny, Bovineone, Dialectric, SmackBot, Gobonobo, Cydebot, Alaibot, KimDabelsteinPetersen, Escarbot, JamesBWatson, Birdbrainscan, VS 78, Spilbrick, Addbot, FluffyWhiteCat, Yobot, AnomieBOT, Citation bot, Tinndel, Guoguo914, Dan Olner, Menwith, Ujjwalsaha1980, Slastic, Citation bot 1, RjwilmsiBot, KLBot2, Bibcode Bot, BattyBot, DeWikiMan, Esuthest, Eborus, Monkbot and Anonymous: 12

- **List of emerging technologies** *Source:* https://en.wikipedia.org/wiki/List_of_emerging_technologies?oldid=688884753 *Contributors:* General Wesc, Ronz, Reddi, Greenrd, Korath, KellyCoinGuy, Vfrickey, Xanzzibar, Akadruid, BigDog, Wolfkeeper, Elias, Gracefool, John Abbe,

## 34.6.2 Images

### 34.6.3  Content license

- Creative Commons Attribution-Share Alike 3.0